Theory and Methods for Supporting High Level Military Decisionmaking

Paul K. Davis, James P. Kahan

Prepared for the United States Air Force

Approved for public release; distribution unlimited

PROJECT AIR FORCE

The research described in this report was sponsored by the United States Air Force under Contracts F49642-01-C-0003 and FA7014-06-C-0001. Further information may be obtained from the Strategic Planning Division, Directorate of Plans, Hq USAF.

Library of Congress Cataloging-in-Publication Data

Davis, Paul K., 1943-
  Theory and methods for supporting high level military decisionmaking / Paul K. Davis, James P. Kahan.
    p. cm.
  Includes bibliographical references.
  ISBN 978-0-8330-4039-8 (pbk. : alk. paper)
  1. Command of troops. 2. Decision making—Methodology. 3. Military planning—Decision making.
  I. Kahan, James P. II. Title.

UB210.D388 2007
355.6'830973—dc22

                                                                                    2007012304

The RAND Corporation is a nonprofit research organization providing objective analysis and effective solutions that address the challenges facing the public and private sectors around the world. RAND's publications do not necessarily reflect the opinions of its research clients and sponsors.

RAND® is a registered trademark.

Published 2007 by the RAND Corporation
1776 Main Street, P.O. Box 2138, Santa Monica, CA 90407-2138
1200 South Hayes Street, Arlington, VA 22202-5050
4570 Fifth Avenue, Suite 600, Pittsburgh, PA 15213-2665
RAND URL: http://www.rand.org/
To order RAND documents or to obtain additional information, contact
Distribution Services: Telephone: (310) 451-7002;
Fax: (310) 451-6915; Email: order@rand.org

Preface

This technical report joins basic concepts in decisionmaking theory with practical methods and tools for useful high-level decision support, such as is needed by a Joint Force Air Component Commander or, in peacetime, an Air Force Chief of Staff doing strategic planning and related resource allocation. The research described here was sponsored primarily by the Air Force Research Laboratory (AFRL) and conducted within the Aerospace Force Development Program of RAND Project AIR FORCE as part of a fiscal year 2005–2006 study, "Capabilities Analysis for Decision Support." Some of the material covered also benefited from cooperation with ongoing RAND Corporation projects sponsored by the Office of the Secretary of Defense. This report builds on prior research on decision support, *Implications of Modern Decision Science for Military Decision-Support Systems*, by Paul K. Davis, Jonathan Kulick, and Michael Egner, MG-360-AF, 2005.

RAND Project AIR FORCE

RAND Project AIR FORCE (PAF), a division of the RAND Corporation, is the U.S. Air Force's federally funded research and development center for studies and analyses. PAF provides the Air Force with independent analyses of policy alternatives affecting the development, employment, combat readiness, and support of current and future aerospace forces. Research is conducted in four programs: Aerospace Force Development; Manpower, Personnel, and Training; Resource Management; and Strategy and Doctrine.

Additional information about PAF is available on our Web site at http://www.rand.org/paf.

Contents

Figures

Tables

Summary

A Framework for High-Level Decision-Support Systems

In this report, we describe an approach to high-level decision support for a Joint Forces Air Component Commander in combat operations or a Chief of Staff in defense planning. Our central theme is the fundamental importance of dealing effectively with uncertainty, whether in effects-based operations, building what the Air Force refers to as the Commander's Predictive Environment, or planning future forces with the methods of capabilities-based planning.

Although many features of the future can be predicted with reasonable confidence, our emphasis is on the many that cannot. Indeed, high-level decisionmakers are commonly afflicted with *deep* uncertainties that materially affect the choice of a course of action but that cannot be substantially resolved by merely working harder. Such issues have been studied in a variety of fields, and the answer to the problem they pose is clear: *The solution is to adopt a course of action that is as flexible, adaptive, and robust as possible.* This is in contrast to making a best estimate of the future and preparing only for that.

Our central idea is that instead of treating uncertainty as an annoyance that merely disrupts developments on the margin, it is better to proceed with the *expectation* of surprise developments and to have great skill in recognizing when adaptations are needed and in making them at the right time. Although this may seem straightforward, it is diametrically different from the attitude reflected in much work on decision-support systems, including the related processes for developing courses of action. That attitude reflects the uncritical "can-do" spirit often reflected in concept development and even in doctrinal writings. As discussed in the Appendix, some official materials on effects-based operations convey this attitude, which undercuts serious thinking. To make matters worse, much related military terminology seems as though it were designed to complicate and obfuscate. A good decision-support environment should have clear structures and terminology; it should confront uncertainty directly.

This report sketches the framework of a high-level decision-support environment consistent with an uncertainty-sensitive approach. That framework includes principles such as being top-down, expressing concepts in simple and intuitive language, dealing explicitly with risk and uncertainty, and providing zoom capability so that decisionmakers can readily discover and question the bases for key assumptions and assessments.

The framework recognizes that high-level decisionmakers vary greatly in style. Some are relatively more "rational-analytic," preferring to see many options and analyses. Others are more intuitive, or "naturalistic," preferring to leap ahead in developing the course of action

xiv Theory and Methods for Supporting High-Level Military Decisionmaking

they believe is suitable, leaving the filling in of details and the validation of assumptions to subordinates. Each style has its own considerable strengths and weaknesses. Although schoolbooks and decision-support-system designers tend to favor rational-analytic methods, and real-world leaders often lean toward intuitive methods, we suggest an approach to top-down decision support that could accommodate both styles, attempting to exploit the strengths and mitigate the weaknesses of each. The product of decisionmaking in both cases should be flexible, adaptive, and robust strategies (FAR strategies); the question is how best to develop and choose among FAR strategies using processes and methods decisionmakers themselves are comfortable with.

The key to our approach is recognizing that (1) typical rational-analytic decisionmakers are aware that the options presented to them may lack creativity, imagination, and an adequate appreciation of "soft" factors; and (2) typical intuitive decisionmakers are aware that (unstated) risks exist in executing their strategies. Our framework attempts to serve both types of decisionmaker—by enriching options and their evaluation for the former, and by identifying potential roadblocks to be overcome in achieving the intent of the latter. Although the processes supporting the two styles will usually be different, in either case the result should improve the courses of action developed and chosen.

Consistent with the broad framework, we describe two explicit methods and their related tools. The first involves *portfolio-style thinking and analysis*, which is currently being used in defense planning and which we adapt here for supporting commanders. It is a good mechanism for balancing risks and other considerations in choosing a course of action.

The second method is a novel modification of what has come to be called *foresight exercises*. This method addresses the need to include humans effectively in dealing with uncertainty. Although it is only one of several methods for doing so, it is interesting and would fill a gap. The intent of this and other methods we mention in less detail is to provide decisionmakers with a more analytically structured and intellectually rigorous approach to human gaming—an approach that preserves the virtues of human gaming while improving upon the validity of conclusions drawn from it. The term *foresight* applies because the purpose is not to "predict" how developments will unfold, but rather to have the foresight to recognize enough of the possibilities to be materially and mentally prepared to adapt when surprising developments occur, as they assuredly will.

Implications for Investment in Decision Tools

Having provided a framework and presented some concrete examples, we next present a more extensive discussion of available methods and enabling technologies and make some recommendations about investment priorities.[1]

The implications for investment are several. Generally, a high priority should be placed on methods for discovering, constructing, and evaluating FAR strategies. If models and simula-

---

[1] This portion of the report draws heavily upon research done cooperatively with projects for the Office of the Secretary of Defense (Dreyer and Davis, 2005; Davis and Henninger, 2007; Davis and Shaver, unpublished).

tions are to be consistent with this priority, they must themselves be adaptive so as to represent appropriately the dynamic and adaptive behaviors of political and military leaders, the organizations they command, societal groups, and so on. Accomplishing this will require adopting concepts from complex adaptive systems and system dynamics, among other methods. The concepts and methods needed should include not only the most common variety of so-called agent-based models, which are developed bottom-up with an eye toward discovering emergent phenomena, but also approaches taken from control theory, game-theoretic simulation, and operations research. Further, more investment should go into top-down approaches, which have the potential of better representing all-important command-and-control considerations, including those for a highly distributed world.

Another implication for investments relates to the need to make better use of people. A sober view of reality quickly reveals that current-day modeling and simulation are nowhere near adequate for thinking through the myriad of relationships and possibilities that arise in human and other complex adaptive systems. In contrast, human-intensive methods such as war-gaming, foresight exercises, Red-teaming, and more-formal activities such as assumption-based planning all have the virtue of unleashing human creativity and drawing on broad-based human expertise and intuition. Typically, however, they are poorly structured and potentially misleading. It follows that those developing decision-support environments should invest in refining such methods and making them more rigorous and in developing approaches to simulation that make optional human participation easy and productive, as in the new activity called *massive scenario generation*, which involves a sequence of human play, model development, scenario generation, observation, and iteration.

One aspect of such investment should be a new round of analytical work to better understand how and when to best use experts—whether to inform model development, consult during crisis or operations, or provide outside critiques. Although many techniques for using experts have been developed over the years, advancements in effects-based operations or development of a Commander's Predictive Environment will require expanding upon them, with special attention to the need to better represent uncertainties and risks meaningfully and to abstract insights for well-hedged planning after having done so.

Acknowledgments

This report builds on previous research sponsored by the Office of the Secretary of Defense and the Air Force Research Laboratory (Rome, New York). Some of the work reported here benefited greatly from collaboration with RAND colleagues James Bonomo, Maarten Botterman, Paul Dreyer, Richard Hillestad, Russell Shaver, and Henry Willis. We thank Kip Miller of RAND and Eugene Santos of Dartmouth College for constructive reviews of an earlier draft of this report.

Acronyms

AT&L	Acquisition, Technology, and Logistics
DIME	diplomatic, information, military, and economic
DoD	Department of Defense
EBO	effects-based operations
FAR	flexible, adaptive, and robust
FFRDC	federally funded research and development center
JFACC	Joint Forces Air and Space Component Commander
JFCOM	Joint Forces Command
JICM	Joint Integrated Contingency Model
JIPB	joint intelligence preparation of the battlefield
JTFC	Joint Task Force Commander
JWARS	Joint Warfare System
OODA	observe, orient, decide, act
OSD	Office of the Secretary of Defense
PAT	Portfolio-Analysis Tool
PAT-MD	Portfolio Analysis Tool for Missile Defense
PMESII	political, military, economic, social, information, and infrastructure
POM	Program Objective Memorandum
RSAS	RAND Strategy Assessment System

CHAPTER ONE

Introduction

Objectives

The purposes of this report are to

- Describe a framework for supporting high-level military decisionmakers in devising and choosing among courses of action;
- Design the framework to account for differences in context and what may be called *analytical* and *intuitive* decisionmaker styles;
- Illustrate some particular analytical methods and tools that can provide this support; and
- Draw implications for next steps in related research and development.

The framework described can be applied both to force planning and to some aspects of operations planning.

We approach these objectives with a particular theme: the fundamental importance and feasibility of dealing effectively with uncertainty. Although many features of the future can be predicted with reasonable confidence, our emphasis is on the many that cannot. Indeed, high-level decisionmakers are commonly afflicted with *deep* uncertainties that materially affect the choice of a course of action but that cannot be substantially resolved by merely working harder. Such issues have been studied in a variety of fields, and the answer to the problem they pose is clear: *The solution is to adopt a course of action that is as flexible, adaptive, and robust as possible.* This is very different from planning for the best-estimate future.

Much of what we discuss in this report has broad applicability, but this particular study was motivated by (1) decisions that might be faced by a Joint Task Force Commander (JTFC) or his Joint Forces Air and Space Component Commander (JFACC) and (2) decisions faced by defense planners (sometimes called force planners), such as the Secretary of Defense or Service chiefs. It was also developed with a specific eye toward problems in which political, military, economic, social, information, and infrastructure factors (so-called PMESII factors) all come into play, and decisionmakers seek to employ a mix of diplomatic, information, military, and economic instruments (so-called DIME instruments). These problems are the norm in modern conflict, whether in contemplating coercive use of airpower in the Balkans during the 1990s

(Clark, 2001; Lambeth, 2001) or combat, counterinsurgency, and stabilization operations in Afghanistan and Iraq (Franks, 2004; Gordon and Trainor, 2006). Such problems are highlighted in the recent *Quadrennial Defense Review* (Rumsfeld, 2006).

Effects-Based Operations and the Commander's Predictive Environment

High-level decisionmakers are responsible for establishing and pursuing suitable visions and objectives. They are also concerned about assuring that these are mapped appropriately into more-concrete objectives (operational objectives) and actions (tasks to be accomplished).[1] The new predoctrinal concept or command philosophy of effects-based operations creates an intermediate between objectives and actions—the *effects* desired as the result of accomplishing the tasks.[2] That is, if operational objectives are still relatively abstract, identifying more-specific objectives and expressing them in the language of effects can be useful. For example, to achieve the objective of defeating enemy air defenses, one subordinate effect that might be sought would be to neutralize central radars. That effect might be accomplished by physical destruction or electronic attack.[3]

Importantly, effects-based-operations methods also convey a strong sense of the system and causal relationships, thereby improving the logic connecting the commander's intent to tasks and laying the groundwork for considering alternative taskings and potentially necessary adaptations.

The implication for decision support is that high-level decisionmakers should be seeing summary materials that focus attention on the relative ability of courses of action to achieve *all* objectives and related effects—not just a subset that is easy for analysts to deal with. This need for comprehensiveness applies to both force planning and operations planning—e.g., to the Chief of Staff of the Air Force or to a JFACC. Providing such a comprehensive view is part of taking a good systems approach and is consistent with a sound approach to effects-based operations.

An effort closely related to effects-based operations is seeking to develop what has been called, unfortunately, the Commander's Predictive Environment.[4] This is the intelligence coun-

---

[1] There is a long history of related methodologies, including the "strategies-to-tasks" framework associated with Glenn Kent (Lt. Gen. USAF, ret.) and other colleagues at RAND (Warner and Kent, 1984; Pirnie and Gardiner, 1996; Kent and Ochmanek, 2003).

[2] Effects-based operations (EBO) has been discussed by a number of authors (Deptula, 2001; McCrabb, 2001; Davis, 2001; Smith, 2003, 2006; Rayburn, 2006; Jobbagy, 2006) and is becoming doctrinally established (U.S. Joint Forces Command, 2006a; Air Combat Command (ACC/A3C), 2005), despite continued controversy due to confusing terminology and a lamentable tendency of proponents to emphasize predictiveness, even where predicting with confidence is not realistically possible. The controversy and its relevance to this report are discussed in the Appendix.

[3] As discussed in the Appendix, it can reasonably be argued that it would be better to refer to subordinate objectives, rather than introducing the language of effects.

[4] The tenets of the Commander's Predictive Environment are described in several recent documents (U.S. Air Force Air Materiel Command, 2006; U.S. Joint Forces Command, 2006a). Some of the issues are analogous to those involved in establishing an Army-oriented commander's information needs (Kahan, Worley, and Stasz, 2000).

terpart to effects-based operations. It, like effects-based operations, relies heavily on a systems perspective that seeks—using increasingly prevalent Pentagon acronyms—to apply the various DIME instruments to problems beset by the many PMESII factors. It envisions proactive intelligence preparation of the battlefield with multiple levels of thinking and seeks to take due account of the near-term future battle space and of enemy capabilities.

Outline of This Report

The report proceeds as follows.[5] Chapter Two discusses uncertainty issues arising in higher-level decisionmaking that are central in developing a decision-support environment. Chapter Three places our work in the context of the current distinctions between rational-analytic and naturalistic approaches in decisionmaking theory. The chapter argues that a synthesis of approaches is not only possible but necessary, and that this is important if decision support is to meet the needs of different decisionmaker styles. Chapter Four describes a framework that moves toward such a synthesis. Chapter Five presents two specific new methods developed at RAND that are consistent with that framework. One adapts portfolio-management techniques to decision support for commanders; the other adapts a gaming technique. Chapter Six then takes a broader view of methods and describes priorities for other methodological developments and related tools. Chapter Seven is a brief wrap-up. The appendix discusses in more detail certain controversies involving effects-based operations.

---

[5] Preliminary versions of some material in this report have been presented at conferences (Davis, 2005a; Davis and Kahan, 2006; Kahan and Davis, 2006). Initial recommendations for research and development were offered in a technical meeting on the Commander's Predictive Environment held in Arlington, Virginia, by the Air Force Research Laboratory, January 31–February 1, 2006. Discussion of portfolio methods draws on work done for the Office of the Secretary of Defense (Davis and Shaver, unpublished); discussion of modeling and simulation draws on work for the Office of the Secretary of Defense (OSD) (Davis and Henninger, 2007).

CHAPTER TWO

Uncertainty, Risk, and Choice

Initial Comments

Virtually all high-level decisionmaking occurs in an environment where not everything that would be nice to know is known. In other words, the environment is uncertain, so any choice involves some degree of risk, sometimes a large degree. The literature on accounting for risk and uncertainty in analysis is voluminous,[1] but in this technical report we focus rather narrowly on high-level military decisionmaking and on extensions of earlier RAND work that we considered particularly relevant. We also refer frequently, although selectively, to military doctrinal materials and to planning methods of current military significance.

In our examination of official materials on effects-based operations and the Air Force's Commander's Predictive Environment (see the Appendix), we found little discussion of uncertainty, except for lip service and in auxiliary documents. As an example, we noted an illustrative vignette carried through the exposition of the Joint Forces Command's *Commander's Handbook* (U.S. Joint Forces Command, 2006a), which begins with certain assumptions and then, when the assumptions are shown not to have been correct, goes on to have the commander merely ask for iteration of the same type of assumptions-sensitive planning (p. IV-17)—as though the next time will "bring the magic." In contrast, the approach recommended throughout this report is to supplement emerging doctrine by confronting directly the many uncertainties in the decision environment and finding ways to deal with them in developing and choosing among courses of action. In this way, the planning recognizes that many assumptions will prove false and builds into the course of action the potential for agile adjustments when the time comes. We shall discuss this approach later in the report where we present an example. The remainder of this chapter provides more background on uncertainty, risk, and choice.

Uncertainty

Even though it is not obvious from current official documents or practice, the *principal* challenges associated with both effects-based operations and the Commander's Predictive

---

[1] A good introduction to risk and uncertainty issues is a book from Carnegie-Mellon University used in many graduate courses (Morgan and Henrion, 1990). Haimes (2004) also discusses such issues, but more specifically in the contexts of systems engineering.

Environment involve uncertainty and risk.[2] High-level planners are often confronted with enormous uncertainties, some of which imply risks and others of which signal potential opportunities. For high-level decisionmakers in particular, many uncertainties are likely to be deep. In our parlance, *deep uncertainty refers to materially important uncertainties that cannot be adequately treated as simple random processes and that cannot realistically be resolved at the time they come into play.* We argue that the deep uncertainty must be acknowledged in the planning process—either by considering alternative courses of action based on the different possible uncertain factors, or by seeking to buy time or information to reduce the uncertainties and have a better understanding of the possible effects of operations. Our definition goes beyond that used by statisticians, i.e., that deep uncertainty exists when one does not know even broad characteristics of the applicable probability distribution.[3] For our purposes, that definition is too narrow and reflects a bias toward mathematical analysis. An example of deep uncertainty relevant to defense planners is imagining the strategy of a future adversary commander in a future war in a future set of circumstances, all of which are hypothetical and unknowable. Interestingly, even a current-day general may have deep uncertainty about his adversary's strategy because of lack of knowledge about the adversary as a person, his history and training, his psyche, and his values. This was the case at the outset of the 2003 Iraq war (Gordon and Trainor, 2006, pp. 182–186). Deep uncertainty is often ubiquitous when behaviors of individuals and groups matter to outcome, as is usual in the domain of political, military, economic, social, information, and infrastructure (PMESII) factors and in attempts to use diplomatic, information, military, and economic (DIME) instruments.[4]

We have emphasized the uncertainty issue for years, both in nondefense contexts (Botterman, Cave, Kahan, Robinson, Shoob, Thomson, and Valeri, 2004) and in defense work that described parametric and probabilistic "scenario-space" methods for confronting uncertainties systematically (Davis, 1994; Davis, Gompert, and Kugler, 1996; Davis, 2002a), methods now incorporated in OSD policy guidance to analysts. Looking across problem domains ranging from football to cybercrime to health care to corporate competition to war, *the primary conclusion is recognition that the answer is the ability to adapt.* When it is not possible to predict the future with a useful level of confidence, success will often depend on adjusting as necessary to what actually develops. This has been discussed for organizations broadly (Light, 2004), and the implications for officer education have also been drawn in connection with networked forces (Gompert, Lachow, and Perkins, 2005) and more generally (Tilson et al., 2005). Although *adaptive* in everyday speech covers the whole territory, we prefer to decompose the broad concept and to differentiate the needs for flexibility, adaptiveness, and robustness in one's

---

[2] See the Appendix for a discussion of the controversy related to effects-based operations, much of which pertains to uncertainty. As discussed in earlier work (Davis, 2001, p. xiv), we believe that uncertainty should be treated as a front-and-center issue, especially when dealing with phenomena involving people's decisions and behavior.

[3] This is sometimes called Knightian uncertainty (Knight, 1921).

[4] The criticality of dealing with the PMESII factors and the DIME instruments is emphasized in recent government policy documents such as the *Quadrennial Defense Review Report* (Rumsfeld, 2006). The challenges of doing so are at the frontiers of modeling and simulation, as discussed in a recent National Academy report (National Research Council, 2006). Traditional modeling and simulation has emphasized what are sometimes called "physics effects," e.g., attrition from weapon exchanges, which is in many ways much easier than understanding and modeling the other PMESII dimensions.

strategies. By our definitions, *flexibility* refers to the ability to perform different missions (e.g., in one region or another, or to go from warfighting into a security and stabilization activity) or different tasks. *Adaptiveness* in this more limited context refers to the ability to adjust readily to diverse circumstances (e.g., political-military context, enemy strategies, warning time). *Robustness* refers to the ability to withstand both foreseen and unforeseen shocks, such as surprise attacks or the loss of an important battle.[5] In this nomenclature, *the solution to uncertainty is to find flexible, adaptive, and robust (FAR) strategies.* This goal is in dramatic contrast to seeking an "optimal" strategy that assumes a particular future, as occurs when people take particular planning scenarios too seriously. Official emphasis on such considerations began to appear in the Department of Defense's (DoD's) first *Report of the Quadrennial Defense Review* (Cohen, 1997) and were at the core of DoD's embrace of capabilities-based planning in 2001 (Rumsfeld, 2001), which has been reinforced subsequently (Rumsfeld, 2006). Capabilities-based planning is particularly salient in an era when U.S. military forces have—for 16 years—been used in a long series of wars, lesser contingencies, and other operations (including the "global war on terrorism" and the effort to stabilize Iraq) that were not anticipated in earlier years. The future looks no more settled than the recent past.

Risk

Defense Department officials addressing force planning are particularly concerned about managing risks[6] (Rumsfeld, 2006), as discussed at length at an April 2006 Military Operations Research Society (MORS) conference on capabilities-based planning (see www.mors.org). One aspect of this has to do with defense planning, such as the acquisition of future forces and capabilities. Another part is related more to operations planning.

Defense Planning

Defense Department officials often refer to the need to identify areas in which "more risk can be taken" (Henry, 2006), but in more classic terminology, they actually have in mind finding bill payers or offsets. It is straightforward for DoD components to identify gaps in capability areas, but it is more painful and difficult to find activities that can be given less funding in order to pay for the gap-filling. The issue, then, is actually the eternal challenge of programming and budgeting (Hitch and McKean, 1965).

---

[5] The three words have diverse meanings. *Flexibility* often is just a synonym for adaptiveness, but our usage is consistent with a common military meaning, analogous to defining it as, say, the ability to shift readily from one campaign objective to another (United States Air Force, 2003, p. 30). Our use of *robust* is consistent with classic literal dictionary definitions referring to the ability to withstand stress (*American Heritage® Dictionary of the English Language*, 4th ed., 2003).

[6] As used here, *risk* is a measure of the potential for bad developments. It can be regarded as a function of the probability of various events and the harm those events would cause. In practice, people and society are often unable to make useful estimates of the probabilities, the extent of harm, or both, and they therefore rely on heuristics such as following "good practices" and avoiding "risky situations" altogether. Some people, of course, are much more risk-tolerant than others. Teenagers are notorious for believing themselves to be immortal. In this report, we do *not* use the definition of risk sometimes used by mathematicians, who distinguish between risk and uncertainty by claiming that the former has a known probability distribution, while the latter does not.

In principle, at least some of the actions needed to balance budgets intelligently may not involve risks, e.g., eliminating waste, fraud, and abuse; canceling an inferior weapon system; or cutting back in "lush" capability areas. Thus, the terminology of *balancing risk* seems odd to those schooled in defense economics and systems analysis. That said, the proponents of most weapon-system programs and force components can identify risks that would indeed be increased by cutbacks, even if DoD critics argue otherwise. As a result, the terminology of balancing risks and being willing to take more risks in some areas to pay for filling capability gaps in others has merit. In any case, it is the terminology currently being used by policymakers. Whatever terminology is used, *a decision-support system should address risk directly and well*. A system that addresses risk in an effective manner will encourage development of FAR strategies.

Operations Planning

Uncertainty and risk in operations are treated by what the Joint Staff calls Adaptive Planning (Bankston and Key, 2006; Hoffman, 2006). Oriented primarily toward peacetime development of operations plans that will be reasonably up-to-date, the approach emphasizes building in branches and thinking in terms of capability packages that can enhance adaptiveness.[7] Over time, this approach should significantly affect both plans and decision-support systems for combatant commanders and component commanders such as JFACCs.

Choice

Even if uncertainties and risks are modest, the necessity of choice will still exist, because of limited resources. In recent years, it has often seemed that DoD has been able to avoid tough choices because the defense budget is currently so large that they are unnecessary. Defense spending, however, is cyclical, and the time is approaching when budget crunches will begin to hurt as bills come due for underfunded programs and the recapitalization of materiel used in operations of the past decade or so. For defense planners, then, decision support should facilitate economic choice.

For commanders, the resource issue is less about budgets than about people and materiel. The pinch is not always severe. During the first Gulf war in 1991, the Balkan conflicts of the 1990s, and the combat phase of the current Iraq war (2003), air forces were more than ample, and the most difficult choices could be avoided. However, the U.S. ground commanders in Iraq today are acutely aware of how limited their resources actually are, and such problems will in the future beset both air-and-space and maritime commanders, especially in situations with limited strategic warning and limited use of foreign bases.

---

[7] Taking such an approach in operations planning was recommended to the Joint Staff's J-5 in a 1993 RAND report (Davis and Finch, 1993), which presaged later recommendations about capabilities-based planning (Davis, 1994; Davis, Gompert, and Kugler, 1996; Davis, 2002a).

Putting It Together: What Decision Support Is Needed?

It follows that a decision-support system should highlight issues of uncertainty, risk, and choice. This must be done hierarchically, at different levels of abstraction, but these issues must not be relegated to mere footnotes, appendices, or background slides. Their essence must rise to the top. This is often not easy, especially given the differing styles of decisionmakers, as discussed in the next chapter.

CHAPTER THREE

Reconciling Different Approaches to Decisionmaking

Background

As discussed in a recent RAND review of modern decisionmaking theory conducted for the Air Force Research Laboratory (Davis, Kulick, and Egner, 2005), it is particularly useful in discussing high-level military decision support to distinguish between the mindsets and methods associated with so-called rational-analytic decisionmaking and those of naturalistic (or intuitive) decisionmaking.[1] These appear to be sharply in conflict (even if the underlying science is not). Analysts and decision-support developers tend toward their approximation of the rational-analytic style of comparing and choosing among multiple options, whereas many real-world high-level leaders are oriented toward intuitive approaches based in part on their experiences and in part on the desire to take decisive action (with adjustments later as necessary). Many leaders of nations, armies, and corporations describe themselves as intuitively inclined, and their decision styles reflect this self-description. Business-school students are often exhorted to worry most about getting to market quickly, with the admonition not to be paralyzed by analysis.

The differences between the prescriptions of rational-analytic decisionmaking theory and the real-world behavior of decisionmakers has been discussed for decades by psychologists and decisionmaking theorists. For many years, however, it was widely believed that the rational-analytic style was obviously right (although limited in applicability because of bounded rationality) and that the task was therefore to help decisionmakers adhere to it, rather than to fall prey to the many well-known cognitive biases. In more-recent years, it has been noted that the intuitive style of decisionmaking is often very effective and should therefore not be scorned. Further, it has increasingly been appreciated that efforts to implement rational-analytic methods often do violence to many of the real issues, much as overly "rational" analysis in our day-to-day world often lacks common sense. Analyses may, for example, (1) suppress

---

[1] The literature on decisionmaking and decisionmaking theory is voluminous. An earlier RAND study led by one of the authors (Davis) provides an extensive, although selective, review of that literature, with the intention of highlighting elements of modern decision science particularly applicable to high-level military decision support. That review included scores of citations relating to, e.g., utility theory, the theory of games, systems analysis, decision analysis, policy analysis, bounded rationality, heuristics and biases, naturalistic decisionmaking, and the new technique of exploratory analysis that has been pioneered at RAND. The most salient paradigms for the current report are those of rational-analytic and naturalistic decisionmaking. The former is more classical and is described in many books (e.g., Keeney and Raiffa, 1993, originally published in 1976). Good references to the latter include two edited books, Klein (1998) and Gigerenzer and Selten (2002).

soft factors that cannot be quantified and analyzed mathematically or measured objectively; (2) ignore considerations of personality and political context; (3) represent only a subset of values, such as those that can be described in economic terms; or (4) make little use of intuitive judgments and hunches, even though these often prove insightful. These shortcomings can, in principle, be mitigated, but they are so prevalent in practice that caution should be exercised before imagining that an allegedly rational-analytic approach to decision support is necessarily good in practice. Furthermore, many high-level decisionmakers are simply not comfortable with the breadth-first, highly analytical approach and prefer—after only modest analysis—to follow their instincts and visions, adjusting as necessary to realities as they are encountered. If decision support is tailored for a different style, how effective can it be? Not very.

To be sure, the virtues of rational-analytic thinking are many and well documented, even in view of the need to be realistic about how well options can be evaluated given the issues of bounded rationality. Further, the more intuitive approaches to decisionmaking may be psychologically attractive, but to paraphrase, it must be said that when "such an approach is good, it can be very, very good, but when it is bad, it can be horrid." It seems clear that decisionmaking and related decision support need to draw on both approaches, both to exploit their separate virtues and to mitigate their separate weaknesses.

Working with Both Rational-Analytic and Naturalistic Styles

With such considerations in mind, we have been contemplating various ways to achieve a synthesis of theories and styles, one that would avoid the most egregious errors of both and would prove more effective in decision-support systems than the pure rational-analytic paradigm. Some aspects of synthesis were identified in earlier work (Davis, Kulick, and Egner, 2005), but we offer new suggestions here, specifically for decision-support systems:

- Assure that risks and risk mitigation are covered effectively, but provide multiple mechanisms for doing so, in order to deal with different decisionmaker styles—in essence, create our own FAR strategies for dealing with risk.
- Provide a rational-analytic decisionmaker with decision support that highlights risks and ways to mitigate them up front, as part of his choice of approaches (a breadth-first approach). In doing so, and as a distinct departure from what has often been done, present alternatives that are not "pure strategies" suitable for particular futures but, rather, FAR strategies.
- Serve a more intuition-driven decisionmaker by developing plans seeking to implement the intended course of action, but—in the process—identify potential hurdles to be overcome and contingencies against which to prepare hedges. In doing so, employ methods such as gaming, Red-teaming, and assumption-based planning (Dewar, 2003)—not to "fight the decision," but to identify what might go wrong and prepare accordingly.
- In both cases, deal with all relevant factors, whether hard and objective or soft and subjective. "Dealing with" may mean creating slots or other opportunities for decisionmakers to use their own judgments, rather than attempting to provide judgments generated by staff on matters beyond their expertise, but the factors should appear naturally, rather than being omitted.

Figure 3.1 summarizes this in a flow chart. Reading it prescriptively, start at the top, center, and decide whether there are recognized strategic dilemmas to be faced. A strategic dilemma exists if different courses of action all have serious risks but uncertainties make it difficult to assess which course is preferable on balance. If there is no strategic dilemma—that is, if the decisions appear to be of a how-to-do-it variety, then proceed along the path on the right side, building a rolling plan with appropriate tactical options. However, insert a process of creative critical review, using, e.g., gaming, Red-teaming, and assumption-based planning, and then revise the plan accordingly so as to be appropriately adaptive and hedged. *Note that this goes far beyond the common practice of assuring that the commander is aware of assumptions underlying the plan.* Even if the commander's head nods, acknowledging risky assumptions, he may not really be better off than before. What matters is doing something to mitigate the risks, and decision support should assist in doing so! Concluding the rightmost flow, if the critical review demonstrates that the plan is bankrupt, then go back to the beginning and take a different approach.

Continuing with Figure 3.1, if strategic dilemmas are recognized at the outset, conduct initial high-level discussions as necessary to make the decision about whether more strategic analysis of alternatives is desired. If not, go with the choice and proceed to build a correspond-

Figure 3.1
A Flow Model for Using Different Approaches in Choosing Options and Building Plans

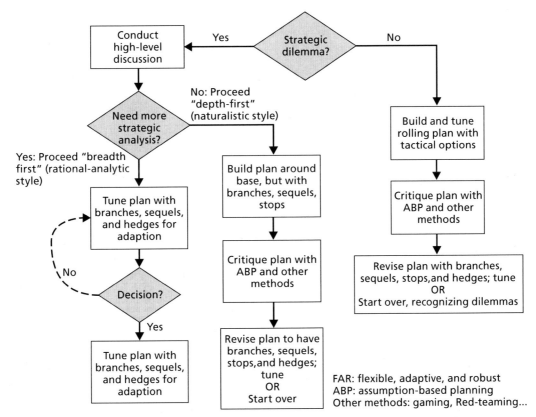

RAND *TR422-3.1*

ing plan, as shown in the central column of Figure 3.1. As with the earlier case, insert a phase of tough critique and revision (or start over). Finally, in the event that strategic dilemmas are recognized and initial discussion concludes that a more careful development of options is necessary, take the path on the left side, a modern version of rational-analytic choice that rejects notions of strategic "optimization" amid massive uncertainty and instead emphasizes development and comparison of alternatives that are each intended to be FAR strategies. In this case, adaptations are built in from the outset and are part of the very fabric of the decisionmaking. The overall intent is that the results of the three processes should all be able to deal with uncertainty reasonably well—not identically, but perhaps well enough in all cases.

A theoretical basis exists for the approach we describe. In particular, empirical research by James March, among others, has found that intuitively driven entrepreneurs are, in fact, interested in risks. They also distinguish sharply between risk-taking and gambling (see, e.g., March, 1994, pp. 35–57). Crane (2006) makes the same distinction in the context of U.S. Army leadership. However, their interest is in identifying risks so that they can be overcome, either by steamrolling over them or by adopting various tactics such as buying out or undercutting competitors. It is interesting to note that while the Bush administration's approach to the war in Iraq clearly reflected strong intuitive considerations, there was a great deal of preparation for things that could go wrong (Woodward, 2004; Franks, 2004). What did go wrong had not been adequately prepared for, as is now well described (Gordon and Trainor, 2006), but our point is that even leaders who had chosen a course of action and did not much like broad discussion of alternatives were quite willing to identify and deal with many of the risks. This point is salient, because one of the great obstacles in developing good decision support has been, in our view, a misunderstanding of decisionmakers and, perhaps, some misplaced technical snobbery by analytically oriented people.

We observe, at this point, a special problem in military organizations: Subordinate officers sometimes "salute a plan" too uncritically, rather than helping the commander identify and avoid problems inherent in the plan. This tendency is far worse with strong-willed commanders who resist suggestions, especially commanders who are intuitively inclined and negative about "paralysis by analysis" and about highly quantitative but ultimately unconvincing analysis of alternatives that misses key factors; or commanders who, in turn, are merely saluting political-level directions. Although this problem certainly exists, we contend that decision support can still very often be effective even with these commanders if the approach described here is adopted.[2]

Figure 3.1 is a simplification, of course. Its purpose is merely to clarify how both rational-analytic and naturalistic styles can be accommodated. In a real process, the courses of action would be assessed, lessons would be learned about which is best for what, new hybrid courses of action would be formed so as to achieve the benefits of two or more of the originals, and so on. Some of that kind of iteration is discussed in Chapter Five, under Introduction to Foresight.

---

[2] One reason for our optimism is that hard-nosed, creative critique of possible strategies is well precedented in the professional military. One good discussion of this is in the Marine Corps' Planning manual (United States Marine Corps, 2000).

Techniques for Creative Critical Review

In the discussion above, we postulated a step that involved creative critical review of a given plan. Such a review should question assumptions, anticipate alternative possibilities, and suggest ways to avoid or mitigate risks. A tenet of using this method is that by confronting uncertainty and the many dimensions of a problem and by considering a broad enough range of possibilities so as to prepare well—whether with branches or hedges, or merely being primed for adaptation—it is possible to do much better than proceeding with blinders on a path that may well prove to be inappropriate to the facts on the ground. It is not that everything can be anticipated or that a more detailed plan can cover every contingency, but that great value is achieved by doing what one can.

A variety of methods exist for preparing for a range of possibilities, although few are codified or routinely used well. The methods include human gaming, Red-teaming, and assumption-based planning. All of these are intensely human activities, and humans are superb—if given the right tools and incentives—at finding flaws in plans. It is no accident that operational commanders have long used war-gaming to evaluate and improve courses of action. However, it is only sometimes done well and very often done poorly—dealing well enough with "comfortable" uncertainties, but less well with respect to possibilities such an adversary's use of so-called asymmetric tactics or the complex interplay of all PMESII factors. Using agent-based models for the latter is a subject of research interest at the moment, but certainly at present we suggest improving the more directly human methods and adapting them for use by commanders.

In Chapter Five, we describe a particular approach to human gaming, foresight exercises, which has proven valuable in thinking through uncertainty-laden problems. Its success depends both on a creative and thoughtful structuring of the problem space (something quite "analytic," even though not computerized or explicitly mathematical) and on integrating the intuitions of different actors (something quite "naturalistic," even though a response to analytically prepared descriptions).

Red-teaming is also a particularly useful method, although it is difficult to accomplish with adequate creativity and breadth and without corruption of the process.[3] The Millennium Challenge 2002 exercise was marred by a now-notorious shunting aside of the Red team when its activities proved troublesome (and regrettably prescient).[4]

Assumption-based planning is an excellent and admirably well documented method, first developed in the early 1990s for the Army Chief of Staff, General Gordon Sullivan, and subsequently refined and applied in a number of RAND studies. It is a systematic way of uncovering hidden or tacit assumptions, discussing their consequences, and then identifying potential ways to prepare for or hedge against their failure (Dewar, 2003).

---

[3] Some literature exists on the merits of Red-teaming, its challenges, and its historical misuse, with penalties being levied in subsequent campaigns (Murray, 2002, 2003; Defense Science Board, 2003b). The term *Rainbow-teaming* might be better than *Red-teaming*, since it is often necessary to consider many different protagonists.

[4] Although this event was widely reported, the most detailed account actually appears in a popular book, *Blink: The Power of Thinking Without Thinking* (Gladwell, 2005). A partial rebuttal to criticism was given by General Peter Pace (Department of Defense, 2002).

Another important method is exploratory analysis (Davis, 2002a). A recent offshoot is exploratory analysis based on models that have been manipulated for *massive scenario generation*—covering as many sources of uncertainty and challenge as participants can conceive.[5] This approach anticipates iteration between human-intensive brainstorming and critique, computer-driven massive scenario generation, and analysis.

How Much Uncertainty Analysis Is Enough?

A Simple Framework

A next issue is how much uncertainty analysis the decisionmaker can deal with effectively, whether it be through rational-analytic processes or, somewhat more indirectly, through the kind of naturalistic approach allowed for on the right-hand side of Figure 3.1. A conclusion from past RAND work is that confronting issues of uncertainty is so difficult that decisionmakers will do well if they consider uncertainty merely to the extent suggested by Figure 3.2, which is notional. It is expressed in the alternative-courses-of-action format of the rational-analytic decisionmaker. For a more intuitive decisionmaker who is willing to tolerate only

Figure 3.2
A Generic Decision Framework

Course of Action	Most Likely Outcome	Best Outcome	Worst Outcome	Net Assessment
1	Good	Good	Bad	?
2	Good	Very Good	Marginal	?
3	Good	Very Good	Very Bad	?

Zoom explains why worst outcome is very bad

Logical Case	Factor	Factor 2	Factor 3	Worst Outcome
1	Good	Good	Good	Marginal
2	Good	Good	Bad	Bad
3	Good	Bad	Good	Bad
4	Good	Bad	Bad	Very Bad
5	Bad	Good	Good	Bad
6	Bad	Good	Bad	Very Bad
7	Bad	Bad	Good	Very Bad
8	Bad	Bad	Bad	Very Bad

RAND *TR422-3.2*

---

[5] Exploratory analysis that varies the parameters of a model is now reasonably well developed, but techniques and tools for assuring that the model has the structural capacity to represent the dimensions of uncertainty are difficult and ill-developed, as discussed in a recent prototype effort by RAND on massive scenario generation (Davis, Bankes, and Egner, 2006).

adjustments that improve the odds of success, the options might be expressed as "the basic plan," Adjustment 1, and Adjustment 2. The decision support here consists of providing staff evaluations of options that estimate most-likely, best-case, and worst-case outcomes. The net assessment is left to the decisionmaker because his own style and various contextual issues outside the scope of the decision-support group will determine whether, for example, he is more or less risk-averse.

The decisionmaker must be able to review the underlying analysis. If, for example, the decisionmaker did not expect the worst outcome for Course of Action 3 to be "very bad," but rather something more like "marginal" or "sort of bad," then the logic should be reviewable by zooming to the lower part of the figure, which provides the rationale—for Course of Action 3—for evaluations of "worst outcome" as a function of several factors. Analogous tables would apply to the other courses of action. This lower table covers the eight logical cases of "good" or "bad" for the three factors. As indicated in the figure, for Course of Action 3 the analysis uses Logical Case 4 as appropriate in the particular circumstances. In this case, both Factors 2 and 3 are assessed as "bad" and the worst outcome is said therefore to be "very bad." The underlying logic for this hypothetical case assumes that two "bads" would lead to a "very bad," regardless of other considerations. If the decisionmaker disagreed, he could intervene. The framework for doing so would be relatively clear. This approach of simple decision and logic tables arose from 1980s work building artificial-intelligence models to represent decisionmakers in computer simulation (Davis and Arquilla, 1992; Davis, 2002b). That work stimulated aspects of subsequent work in, e.g., influence-net research (Rosen and Smith, 1996). In more recent years, historical research on national crisis decisionmaking has tended to confirm that human decisionmakers strive for something at roughly the complexity of the top portion of Figure 3.2 (unpublished work by Michael Egner and Paul Davis).

Weighting the Factors

There is no indication in Figure 3.2 of how the various columns are to be combined. That is deliberate, because by and large, we believe that decisionmakers should see the factors and do their own integration, especially because staff analysis on this crucial matter is often not synchronized with the decisionmaker's thinking until late in the process. This said, in cases where linear weighted sums make sense (as postulated for the top part of Figure 3.2, but not the lower part), it is easy to insert a "slot" with weights on the three cases, perhaps with straw-man weights written in as shown in Tables 3.1 and 3.2.[6]

Tables 3.1 and 3.2 present illustrative conservative (risk-avoiding) and ambitious (risk-taking) assessments, respectively. The weights in Table 3.1 might apply if the decisionmaker were primarily worried about downside considerations, rather than upsides. The weights in Table 3.2 might apply if the decisionmaker were particularly interested in a glorious outcome (a common characteristic of history's best-known "great commanders").

---

[6] The weightings here relate not to subjective probability, since these weightings are used in reaching the assessments in the first place, but rather to the relative significance placed by the decisionmaker on gains and losses. The personal weighting scheme is a function of decisionmaker personality, current context, and history (Kahneman and Tversky, 1979; Kahneman, 2002).

Table 3.1
A Generic Framework for Decisionmaking Amid Uncertainty, with a Conservative Style Assumed

Course of Action	Most Likely Outcome	Best Outcome	Worst Outcome	Net Assessment
Weights	0.5	0	0.5	
1	Good	Good	Good	Good
2	Good	Very good	Marginal	Marginal
3	Good	Very good	Very bad	bad

NOTE: Based on mapping very bad, bad, marginal, good, and very good into 1, 3, 5, 7, and 9, respectively; using linear-weighted sums; and averaging down on the margin, consistent with the particular decisionmaker's conservatism.

Table 3.2
A Generic Framework for Decisionmaking Amid Uncertainty, with a More-Ambitious Style Assumed

Course of Action	Most Likely Outcome	Best Outcome	Worst Outcome	Net Assessment
Weights	0.5	0.5	0	
1	Good	Good	Good	Good
2	Good	Very good	Marginal	Very good
3	Good	Very good	Very bad	Very good

NOTE: Based on mapping very bad, bad, marginal, good, and very good into 1, 3, 5, 7, and 9, respectively; using linear-weighted sums; and averaging up on the margin consistent with the particular decisionmaker's ambitiousness as expressed by the weights.

The structuring of the problem in Figure 3.1 and Tables 3.1 and 3.2 is different from that in normal decision theory. The word *utility* does not appear, although it is something that can be inferred after decisions are made (Davis and Arquilla, 1992). A decisionmaker who tilts toward weighing the upside beyond what is justified by estimates of probability apparently sees a positive utility for the upside outcome that is larger than a negative utility for the downside.[7]

Significantly, judgments about probability, which are already folded into the assessments, are often correlated with ambitions (related to "motivated bias") and judgments about the ability to shape events by one's own actions, such as rapid decisive operations. That style of thinking has been characteristic of many "great leaders," for better or worse. Napoleon, despite many glorious victories, returned from his winter assault on Russia with only a small fraction of his army still alive. It is not that Napoleon was "wrong" in seeking a glorious outcome; rather, he—like many before and after him—overestimated the plausibility of that glorious outcome and did not pay adequate heed to the risk of catastrophic defeat. That understates, of course, a historically colossal blunder. Those tasked with supporting decisions should bear this in mind and attempt to assist the decisionmaker in disentangling notions about probability and personal values.

---

[7] In standard decision analysis, the goodness of alternatives is measured by "utility," which can sometimes be as simple as expected profit but can also be much more multifaceted and, to some extent, a reflection of the decisionmaker's personal preferences. See, for example, Raiffa's classic introduction to decision analysis (Raiffa, 1968).

Assessing Confidence Level

A decision aid like that described above is useful, but it is far from foolproof. In particular, people are notorious for having greater confidence than is warranted empirically (Gilovich, Griffin and Kahneman, 2002). Even experts will often say that they are 95 percent sure of something (or that "it's a slam-dunk") when their knowledge would better justify a figure of something like 80 percent, or even much less. With this in mind, we suggest that decision support should routinely include something like Table 3.3, which assesses the credibility of estimated confidence levels as a function of process. If the assessments are based only on in-group judgments (including reach-back from the battlefield), a "one-sigma" notion about best- and worst-case outcomes might be moderately credible, but any claim that the outcomes shown were "two-sigma" values (i.e., having less than a 10 percent chance of any worse outcome) should be given low credibility. If, however, the judgments also reflect serious use of Delphi (Linstone and Turoff, 2002) or related methods, drawing upon an outside community of experts (e.g., drawing on the so-called wisdom of crowds, to cite the current popular work on the subject (Surowiecki, 2005)), then the judgment might have moderate credibility. If, in addition, historical evidence supports the judgment, credibility might be considered high. As an example, had the credibility of the assessment of no stabilization-and-reconstruction phase been tested in this way before the Iraq war, it would assuredly have been found to be low. Many general officers were skeptical, and historical work suggested the need for many more forces than were being sent (Gordon and Trainor, 2006; Dobbins et al., 2003). Although legitimate reasons exist for estimating a quick end to conflict, confidence in that estimate should have been deemed low. Related techniques for debiasing confidence have been discussed elsewhere over the years, as in Chapter 9 of Bazerman's *Managerial Decision Making* (Bazerman, 1986).

Table 3.3
Credibility of Claimed Confidence Level in Assessment of a Worst-Case (or Best-Case) Outcome

	Claimed Confidence Level	
Basis of Judgment	Moderate (66%, or "one-sigma")	High (95%, or "two-sigma")
Local staff and limited reach-back to like-minded colleagues	Moderate	Low
Above, plus self-critical reach-back and analysis to gain wisdom-of-crowd benefits	High	Moderate
Above, plus use of empirical data from roughly analogous situations	High	High

CHAPTER FOUR

A Framework for High-Level Decision Support

Against this background, let us now describe what we see as the requirements for a decision-support framework. We discuss, in turn, (1) option development and (2) a portfolio-management approach to option evaluation. Under the latter, we discuss a top-down mechanism of evaluation, treatment of risk, cost-effectiveness analysis, and the mathematics and logic of aggregation. Such matters are important in all applications of the portfolio-management approach, which is ordinarily used for strategic planning, rather than operations.

"Serious" Options, Not Stereotypes

In quantum mechanics and mathematics, defining the orthogonal variables of a problem space is powerful and illuminating. However, in the world of strategic decisionmaking, using orthogonal strategies as options can be positively mischievous—i.e., causing people to think of pure options as legitimate choices when in reality they are impractical. It is sometimes legitimate to use orthogonal dimensions to define a "scenario space" or a space of possible strategies, but even then, the wisest strategies will typically be not "pure," but mixed. There are at least two reasons for this. First, strategic decisionmakers often have multiple objectives; thus, the intent should be to find strategies that do reasonably well for all of the objectives. Second, as discussed in earlier chapters, the generic solution to dealing with profound uncertainty is flexibility, adaptiveness, and robustness of strategy; this generic solution virtually necessitates combinations of variables.[1]

A counterargument is that it is good to use orthogonal options in a rational-analytic process so as to sharpen the issues. Once the choice is made, the option can then be adjusted to be less pure but better hedged. In our experience, however, presenting orthogonal options sometimes encourages apparent single-mindedness, which is then overinterpreted by staff who try to tune plan details so as to be consistent with the boss's intent, thereby deleting or deemphasizing important hedges. Far better, we believe, is decision support that emphasizes a portfolio perspective from the very outset.

---

[1] The emphasis on FAR strategies was accepted by a recent National Academy panel (National Research Council, 2006). See also Lempert, Popper, and Bankes (2003) on robust adaptive strategies, which emphasizes the same idea, but with different language.

The parallels in operations planning might include adopting a strategy that seeks to accomplish multiple goals and that has two or more tacks so as to improve the probability of success. This might include heavy preparation of the battle space with air power and ample ground-maneuver forces *and* information operations *and* making every effort to avoid unnecessary collateral damage. The emphasis on these would be uneven, but all would be included. This would not be a pure strategy.

An example in actual operations was the air war in 1991's Desert Storm. Some officers advocated a pure strategy based on Colonel John Warden's concept of winning the war by attacking concentric circles of power from the inside out. In early versions, the concept saw the Iraqi army, even its Republican Guard, as only minimally important: The "centers of gravity" were seen to be command and control and other natural objects for air attack. A quick and painless victory was predicted for the inside-out strategy. Chairman of the Joint Chiefs Colin Powell was not impressed, nor were others, including the air-component commander (General Horner), who ultimately took a much less pure approach. The actual air attack was quite broad and included a devastating attack of the ground forces seen by others as the center of gravity. This proved wise, since the strategic bombing did not accomplish the intended effects. Air power *was* largely decisive, but for other reasons.[2]

The first leg of the framework, then, is an attitude of emphasizing alternative ways to achieve FAR strategies from the outset. For this, we adapt work on portfolio-management techniques.

Portfolio-Management Tools

General Requirements

As mentioned earlier, the second part of our suggested approach is to use portfolio-management methods and tools[3] to evaluate candidate strategies for their "FARness." Our thinking on this matter stems from considerable work over the past decade, primarily on defense planning (Davis, Gompert, and Kugler, 1996; Davis, 2002a; Hillestad and Davis, 1998; Dreyer and Davis, 2005). In recent work for the Under Secretary of Defense for Acquisition, Technology, and Logistics (USD AT&L), we have identified key elements of a generic analytical framework for reviewing capabilities both in and across capability areas and for making

---

[2] Discussions can be found in a number of sources (Keaney and Cohen, 1993; Lambeth, 1992; Beagle, 2000, pp. 72–78; Pape, 1996, p. 221).

[3] In a portfolio-management approach, investments are made in a collection of categories to achieve balance among conflicting objectives. For example, a financial investor may own stocks, bonds, real estate, and commodities to strike a balance among the desire for long-term capital gain, current income, hedging against inflation, and hedging against fluctuations (see, e.g., Fabozzi and Markowitz, 2002). In defense planning, objectives are more complex and the empirical basis for assessing the likelihood of various developments is much poorer, but a portfolio might involve activities to pursue U.S. objectives in different regions, to maintain various types of military capability, and/or to avoid risks of different sorts. In the portfolio approach, setting priorities is less the point than adjusting the weights of effort within the portfolio. All categories may be important, but circumstances and likely returns on investments call for more investments in some categories than in others (Davis, Gompert, and Kugler, 1996; Davis, 2002a). The relationship between defense applications and finance is more metaphorical than mathematical.

economic trade-offs among options (Davis and Shaver, unpublished). Those key elements, which overlap with the discussion earlier in this report, are

- Routine use of portfolio-management tools
- Candid, comprehensive assessment of critical-component capabilities, costs, and benefit-cost ratios (for the near, mid, and long term, anticipating strategic adaptations by both adversaries and ourselves)
- Portfolio adjustment to fill gaps, balance risks and opportunities, prioritize by packages, and conduct marginal or chunky marginal analysis
- Two or more levels of zoom where needed
- Clear basis for assessments, whether technical or subjective
- Routinely rigorous analysis, even where subjective
- Parametric capability models for comprehensive analysis
- Families of models, games, experiments, and historical analyses.

In this chapter, we focus primarily on peacetime applications, such as defense planning; in Chapter Five, we illustrate how similar methods can be used to support commanders such as JFACCs.

A crucial objective is to provide a strategic assessment that is comprehensible at a glance, yet comprehensive in touching upon all the critical components, a core concept of capabilities-based planning and system engineering. One key feature is candor—showing aspects of both failure and success. Routinely achieving candor is often difficult, especially in environments characterized by obsession with consensus or with protecting favored programs. Success depends on the environment established by leadership. Staffs can do excellent work, or they can be marvels of obfuscation. The problem is less severe in that regard in operations planning.

The portfolio-management tools should make it easy not only to see gaps, but also to help decisionmakers decide how to adjust the portfolio so as to fill the gaps, balance risks and opportunities, prioritize by groups rather than by discrete activities, and even to conduct investment analysis, such as marginal or chunky marginal analysis.

Such ambitions cannot be achieved with a single level of detail. The top-level summary (e.g., stoplight charts) is insufficient. Decisionmakers must insist on levels of zoom or drill-down. Without this, there is little basis for understanding or for knowing how best to question and adjust assumptions. Although briefings commonly have some *illustrative* detail, allowing for more ad hoc zooms is important. In practice, decisionmakers themselves will do only limited zooming, perhaps as part of some spot-checking, but the results will strongly affect their confidence in the quality of the staff work and programs. The expectation of such spot-checking, in turn, will greatly enhance the rigor of staff work and the clarity with which messages are communicated.

The analytic framework for such assessments should depend on not just one or two models, but on entire families of models, human war games, experiments, and other forms of analysis, including historical studies (Davis, 2005a). Commissioning the assembly of such families, whether centralized or virtual, should be an explicit management action. Success will require drawing heavily on cutting-edge theory and practice in multiresolution modeling and simulation (Davis, Bigelow, and McEver, 2001). The idea of families of tools has been discussed extensively in several forums, including recent efforts to support development of a Modeling and Simulation Master Plan for analysis (Davis and Henninger, 2007) and a National Academy study on DoD's modeling and simulation (National Research Council, 2006).

Treatment of Risk Within a Portfolio-Management Decision-Support System

The next question is how to represent risk within a portfolio-management decision-support system. Although a purely generic taxonomy of risks has eluded us because so many differences exist among application areas, the following example has proven useful in current work:

- Acquisition risks
  - Feasibility
  - Programmatic (cost, schedule)
  - Political stability of support along the way
- At-the time strategic risks
  - Foreign and domestic support (e.g., congressional support, allied permission to use bases)
  - Warning and decision time
- Operational
  - Effectiveness in achieving the principal effect sought
  - Control of other effects (e.g., collateral damage, perceptions, behaviors)
- Subsequent strategic-effect risks (e.g., the risk that a coalition will disintegrate, that domestic support will weaken).

The taxonomy is organized chronologically. It includes risks involved in acquiring the capabilities in the first place, risks associated with their usability when needed in crisis or conflict, operational risks when actually employed, and risks associated with negative strategic effects (e.g., international perceptions) even if the operation itself is successful and achieves the desired operational-level effects.

If an operational commander, such as a JFACC, were considering options in wartime, a similar taxonomy would apply, but it would not include acquisition risks and it would include increasingly detailed sublevels of operational risk. If, for example, courses of action in a crisis included a possible strike operation deep into another country in retaliation for a that country's support of terrorist operations, the relevant commanders would be concerned about operational risks such as the risk of losing aircraft and pilots in ingress and egress and the risk of not successfully finding and destroying assigned targets. They would also be concerned about at-the-time strategic risks related to allied permissions for basing and overflight and possible political support. Other risks would be considered for the longer run, such as the risk that the strike operations would cause coalition problems. Thus, in evaluating courses of action, they could have categories like those in a portfolio approach for various objectives and various types of risk-avoidance.

Places to Reflect Risk in a Portfolio-Management Decision-Support System

From a technical perspective, the next issue is where and how to represent the various risks in a portfolio-oriented decision-support system. The issue is nontrivial because the basic concept is to proceed top-down and achieve comprehensibility. Given the number of important risks, it is easy to overwhelm the decisionmaker with numerous complex tables and graphics. Although learning how to summarize risks is an ongoing activity of discovery, we have identified the following principal mechanisms (Davis and Shaver, unpublished):

- Measure effectiveness in "bad cases" (e.g., have separate measures of effectiveness for both nominal and worse-than-expected cases)
- Use the graphic equivalent of cautionary "footnotes"
- Include explicit measures of composite risk
  – Inherent risk (typically of a mission rather than a particular option)
  – An option's risk, relative to the inherent risk as a baseline
- Perform underlying calculations of effectiveness with safe-sided assumptions.

Lowest-Level Explanations

The next technical issue with which we have grappled is how to provide immediate explanations of analytical results from detailed technical calculations. We have referred above to zooms, but what is possible and appropriate depends on the nature of the issue. A full discussion of this problem goes far beyond the scope of the present report, but a few points are appropriate here:

- To the maximum extent possible, essential documentation of both assumptions and logic should be included within the decision-support system and should be understandable to the decisionmaker's top analysts.

This can usually be accomplished adequately with a combination of intuitively named variables, well-structured assumption lists, some overview graphics such as "live" exploratory-analysis charts allowing interactive response to questions, and simple logic tables.

Only an accumulation of case histories will allow us to measure the correctness of our judgments, but our initial efforts have been reasonably successful, although they have required substantial effort.

The Mathematics and Logic of Aggregation

One of the most difficult technical issues in designing top-down decision support is summarizing (abstracting) lower-level knowledge in high-level displays. These matters go beyond the scope of this report, but some highlights of our current thinking are as follows (Dreyer and Davis, 2005; Davis, Bonomo, Willis, and Dreyer, 2005):

- Linear-weighted sums are sometimes surprisingly good aggregation mechanisms for decision-support information (Dawes, 1979), but they can also be quite bad. One important case where they are inappropriate is when a system's capability is being assessed and that system depends on each of several critical components performing adequately, i.e., if any one fails, the system fails. A linear approximation implying that one could compensate for failure of one component by buying more of another would be seriously misleading.

- Once linearity has been rejected, it is necessary to allow for several types of nonlinearity. These include (1) thresholds, so that variables are treated as having zero effect if their raw values fall below minimum levels; (2) critical-component effects that cause aggregations of effectiveness to be zero if any critical-component capability is inadequate; and (3) probability-related mathematics.

Cost-Benefit Information and Chunky Marginal Analysis

The last item in our requirements for features in a portfolio-management tool concerns cost-benefit calculations and the related issues of marginal and chunky marginal analysis. We believe the most important points are the following:

- Since any composite measure of overall effectiveness will be suspect because of uncertainties or disagreements about how to aggregate, a mechanism is needed for exploring the consequences of different perspectives about, e.g., the relative importance of different missions and constraints, the relative probabilities of various risks, and so on. Single-assumption-set analysis of effectiveness and cost-effectiveness will have little influence on savvy decisionmakers.

- There is need for both marginal analysis, identifying where to spend or cut the marginal dollar (or billion dollars), and a more chunky type of analysis that uses larger increments of spending (or cuts) and is able to account for S-shaped phenomena such as high-payoff systems that require a huge up-front investment and time before any payoff is achieved.

- There is also a need to be able to do cost-benefit comparisons on large composite options, such as alternative defense programs or alternative POMs (Program Objective Memoranda, the yearly Service program submissions). This is not marginal analysis, but a more strategic level of cost-benefit work. The classes should be related, however, in that the composite options examined seriously should be motivated by more-microscopic analysis; they should be the best-of-breed options for alternative strategies.

Exploratory Analysis

As mentioned above, RAND's portfolio-management tools are based on spreadsheet technology, with related advantages and disadvantages. We find the tools to be good for providing a top-down perspective, but not for going into much depth. The spreadsheet displays are abstractions from work of a more classic systems-analysis variety that includes numerous parametric displays, allowing one to understand how model outputs vary with assumptions. A key element of this in its modern version is *exploratory analysis*, in which all of the key parameters of the problem are varied simultaneously so that one can understand results as a function of those parameters in the complex n-dimensional space. That is, one sees how the parameters interact. For example, a short warning time *and* an unanticipated enemy attack on a vulnerable key system *and* poor command and control may generate a catastrophe, whereas any one of these alone might not.[4]

---

[4] We have discussed exploratory-analysis methods in some detail elsewhere (Davis, 2002a; Davis, Bigelow, and McEver, 2001; Davis, McEver, and Wilson, 2002) and do not repeat them in this report.

CHAPTER FIVE

Examples of New Tools

In this chapter, we illustrate some methods and tools consistent with the spirit of the discussion above. These relate to portfolio analysis adapted to a commander's needs and to a form of gaming called *foresight exercises*. The first of these is relatively analytical and draws upon modeling and simulation (as well as subjective judgments); the second exploits the strengths of human creativity and knowledge but does so in an unusually structured manner. Later, in Chapter Six, we go into somewhat more detail about the methods and tools needed to support JFACCs and other commanders, but we begin here with the particular suggestions, both of which are new for this application.

Portfolio Analysis for a Commander

Introduction

The Portfolio-Analysis Tool (PAT) is a generic offspring of a tool called PAT-MD, developed for the specific purposes of the Missile Defense Agency (Dreyer and Davis, 2005). It was motivated by earlier work with a similar tool called DynaRank (Hillestad and Davis, 1998), which has also evolved over time, in part as the result of military, environmental, and health applications, and in part as the result of lessons drawn from work with PAT and PAT-MD. We shall focus here on PAT.

In a sense, PAT is simply a spreadsheet application, built in the ubiquitous Microsoft Excel®. In practice, such a tool stems from a great deal of thinking, programming, experience, and iteration. As is the case with much software, usefulness depends upon details. In this report, however, we focus primarily on major functionalities.

Figure 5.1 shows a top-level summary chart, simplified from a hypothetical PAT application to operations planning, assessing three options against a number of criteria. The figure is just a familiar scorecard, although it appears in grayscale here, rather than the usual color-coded fashion. Note that the scorecard includes a "risk" column. For simplicity in this particular example, there is no "net assessment" column. The context is that a JFACC is considering three courses of action before recommending an approach to the Joint Task Force Commander (JTFC). The first is the JTFC's clearly preferred option. The commander's sense—based on long experience and intuition—is that the operation must be as coalitional as possible to build cooperation, trust, and shared responsibility. The base plan is to do everything with the coalition partners

Figure 5.1
A Summary Portfolio Display for a JFACC

Measure Course of action	Disarm Enemy Air Forces and Air Defense Detail	Effectively Support Ground Maneuver Forces Detail	Minimize Coalition Losses Detail	Control Potentially Negative Effects Detail	 Detail	Risk Detail
Full togetherness						
Base, but with U.S.-only initial air, missile, and IW strike						
Base, but with U.S.-only initial missile and IW strike						
Color code						
	0.8 to 1.0	0.6 to 0.8	0.4 to 0.6	0.2 to 0.4	0.0 to 0.2	

NOTE: The blank column is simply a visual divider. IW = information warfare.
RAND *TR422-5.1*

involved as much as possible in all activities. The JFACC, however, is concerned that the base plan involves substantial risks. First, the plausibility of destroying the enemy's air forces and air defenses may well depend on achieving surprise, which will be much more risky in a fully coalitional attack. Second, the JFACC is not confident of the prowess of all of the coalitional members' operators or of the reliability of their systems. Although the JFACC has no such concerns about the allies with whom he has worked previously, assigning important functions to others creates worries.

With this background, the second and third options are mere "variants" of the base plan: There is no attempt to fight the JTFC's guiding principle; rather, these options attempt to mitigate the risks of the base plan. Option 2 would do that by having the initial strike conducted entirely by U.S. stealthy aircraft and missiles to achieve maximum surprise. There would be no "local" indicators that a strike was under way. Although coalitional forces would be reasonably ready for battle, they would not be visibly preparing for *imminent* battle. Option 2 is deemed likely to disarm the enemy's air forces and air defense systems, making it possible for friendly air forces to provide superb cover for subsequent ground-force maneuver. Option 3 emphasizes use of only missiles and information-warfare assets during the initial strike. Proponents of this variant argue that it would minimize the possibility of coalitional losses and should be able to disarm the enemy's air forces and air defenses. The idea in both Options 2 and 3 is that there would be extensive consultation and coordination with coalition members, consistent with the JTFC's intent, except that there would be no pre-notification of precisely when the initial strike would occur, to maximize surprise. Proponents of these options believe that allies would probably be assuaged by the consultation and coordination.

Figure 5.1 does not show an "upside" to balance the "downside" (risk), because the entire operation is already being planned for the upside case of decisive victory. The first four columns of Figure 5.1 show four measures of goodness: prospects for (1) disarming the enemy's air forces and air defense; (2) supporting ground-maneuver forces; (3) minimizing coalitional losses; and (4) controlling potentially negative effects, such as fracturing of the coalition due to non-cooperation or strategic blowback due to excessive collateral damage. By these standards, the basic full togetherness plan fares well in avoiding negative effects (white is very good, analogous to green in a stoplight chart). It does less well in the other categories. The variants with a surprise U.S.-only or U.S.-dominated initial strike do very well, except that some tension with alliance members could be anticipated, which might have repercussions, such as delays in ground-force maneuver or non-cooperation of various types. Such problems are assessed to be not very serious (light gray is good, although not very good).

The risk column shows more stark differences. All of the courses of action have risks, but those of the base plan and the second variant are significantly greater. At a glance, the chart suggests that the middle option is preferred.

Figure 5.2 presents a zoom on the risk column, showing risk as having been judged from operational-risk and strategic-risk components. Operational risk is deemed quite high for the baseline option, but low for the first variant (Option 2). In contrast, Option 2 has some strategic risk that the baseline option does not have. Why? Because despite the assessments of the JFACC's staff, the possibility exists that important coalition members will be much more angered than expected when a surprise initial strike occurs about which they were not notified in advance. Although the groundwork would have been laid in advance so that only the timing would be a surprise, perceptions on such matters are not entirely controllable. Even if military-to-military communications were good, reactions of national capitals might be surprising. Thus, there are tradeoffs. Option 3, depending strictly on missiles and information warfare, is deemed to be even more inherently risky. Its operational success would depend on the quality of detailed technical intelligence, always a worry; also, it is inherently difficult to predict perceptions of the option's appropriateness by the world. Temporary effects of cyber-attacks, for example, might be perceived as far more draconian and permanent than they would actually be.

Figure 5.2
A Zoom to Elaborate on the Evaluation of Risk

Level 2 Measure	Operational Risk	Strategic Risk	Warning	Net Score
Course of action				
Full togetherness	■	0.9		0.53
Base, but with U.S.-only initial air, missile, and IW strike	1	0.5		.75
Base, but with U.S.-only initial missile and IW strike	0.4	0.3		0.53

Again, by "eyeballing" Figure 5.2, one can better understand the judgments appearing in the risk column of Figure 5.1.[1] Figure 5.3 shows another zoom, this time of operational risk. It asserts that none of the options involve very much risk in getting to the targets, but the ability to attack the targets successfully and disarm the air forces and air defenses is very much in doubt for the baseline option (because surprise might not be achieved) and somewhat in doubt for the missile-and-IW option because of its dependence on details of technical intelligence and the decreased flexibility of missiles as new information becomes available. This information is not explicit (it would be documented in the PAT spreadsheet) but would be easily explained; the displays are merely devices to assist in the story-telling.

Although this example has been highly simplified, we hope that it conveys the sense of the tool. Actual applications involve more variables, more subtleties, and a good deal of work.

"Zooming" into Underlying Systems Analysis

The zooms discussed above work rather well as long as the issues can be understood in logical terms, as depicted by a table with variables having discrete values, such as low or high, good or bad. A different type of zoom is needed to understand issues in more detail. One natural format for such detailed material is the kind of parametric chart that systems analysts and operations researchers like. Such charts, with overlays, can explain the judgments reflected at higher levels with discrete cases and can do so in a broader context that reveals, e.g., phenomena such as sharp boundaries versus asymptotic tails. We do not show this type of zoom here, but we use it routinely with PAT, sometimes using analytical models embedded in PAT itself and sometimes by pulling up another model running at the same time as PAT.

Marginal Analysis

PAT has been used primarily for investment analysis. In our applications, the options have not been operational courses of action, but rather alternative investment programs. Figure 5.4 presents one output from such an analysis. Each point on the scatter plot of effectiveness versus

Figure 5.3
A Second Zoom Elaborating on the Assessment of Operational Risk

Level 3 Measure	Getting to Target	Disarming Target	Operational Risk
Course of action			
Full togetherness	0.8		
Base, but with U.S.-only initial air, missile, and IW strike	1	1	1
Base, but with U.S.-only initial missile and IW strike	1	0.4	0.4

RAND *TR422-5.3*

---

[1] The warning column is a placeholder for text comments, of which there are none in the example.

Figure 5.4
An Illustrative Cost-Effectiveness Chart for Resource-Allocation Decisions

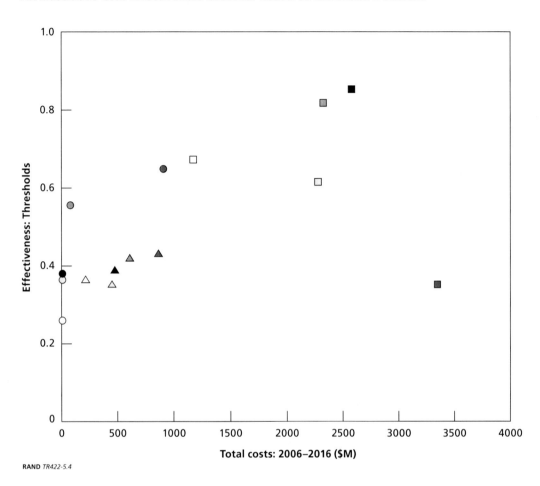

expenditure is a different option. In interactive work, the analyst or briefer "mouses" over a point, at which time a label pops up naming the option. As the example illustrates, by arraying effectiveness versus cost this way, it is easy to see where to spend the marginal dollar (or billions of dollars)—if the composite measure of effectiveness is satisfactory. Since any one such measure is in fact unlikely to be satisfactory, we explore the significance of different perspectives of how to calculate effectiveness. Often, some options are robustly preferable because they have noncontroversial value and don't cost very much—although they may raise organizational hackles. In other cases, relative cost-effectiveness depends sensitively on perspective, and analysis cannot resolve the disagreement. Policymakers must the draw on other considerations to make choices. They may, for example, assess subjectively the relative credibility of options on the basis of their advocates' past performances.

Analysis such as that in Figure 5.4 avoids the long-standing but foolish debate about which comes first, strategy or budget. The plot is "technical." It tells us what is preferred as a function of the money that may be available. And by going into details as to what constitutes effectiveness, one can get a good sense of value. The decision about budget level, then, is often determined by a combination of how much would be obtained with a larger budget and how much less would be obtained with a smaller budget.

An Application of Foresight Methodology to Commanders' Decision Planning

Let us now turn to a completely different kind of tool, one involving gaming in the form of foresight exercises.

Introduction to Foresight

Despite the unfortunate use of *predictive* in the title of the Commander's Predictive Environment, we view the intent of the program as an attempt to establish *foresight*, the definition of which is the ability to envision possible future problems or obstacles (Encarta World English Dictionary). The essence of what are coming to be called foresight methods (Botterman et al., 2004) is identifying the major dimensions of the future that may influence the world and for which establishing the right courses of action at the right time may make a difference. Although the development of foresight methods first occurred in nonmilitary applications, the central ideas are part of an ongoing interaction between military and nonmilitary thinking.

What Is the Foresight Approach?

The foresight approach characteristically seeks the potential drivers of change relative to a simple extrapolation. Because the future is inherently uncertain and multidimensional, planning based on such an extrapolation, or on any one or a few notions about the future, will not do the job (Davis, 2001; Botterman et al., 2004; Lempert, Popper, and Bankes, 2003; van de Riet, van het Loo, and Kahan, 2005). The drivers of change are rarely fully controllable. The changes to be understood may be almost continuous, each so small as to be barely perceived, or they may be discrete events. They may be natural, purposive, or by-products of other purposes.

The foresight approach for our military application constructs potential courses of action in an attempt to achieve desirable futures—i.e., futures with the potential good features we seek and without the undesirable aspects we fear.

The particular foresight approach we have used employs a conceptual framework called XLRM, which was developed recently at RAND (Lempert, Popper, and Bankes, 2003).[2] The concept is sketched in Figure 5.5. Although its original purpose was to illustrate a tool for understanding possible futures 100 years out with a great many scenarios having been generated, the concept is also useful for shorter-term futures and/or small numbers of scenarios.

---

[2] The letters XLRM represent eXogenous factors, policy Levers, Relationships, and Measures.

Figure 5.5
The XLRM Framework for Foresight

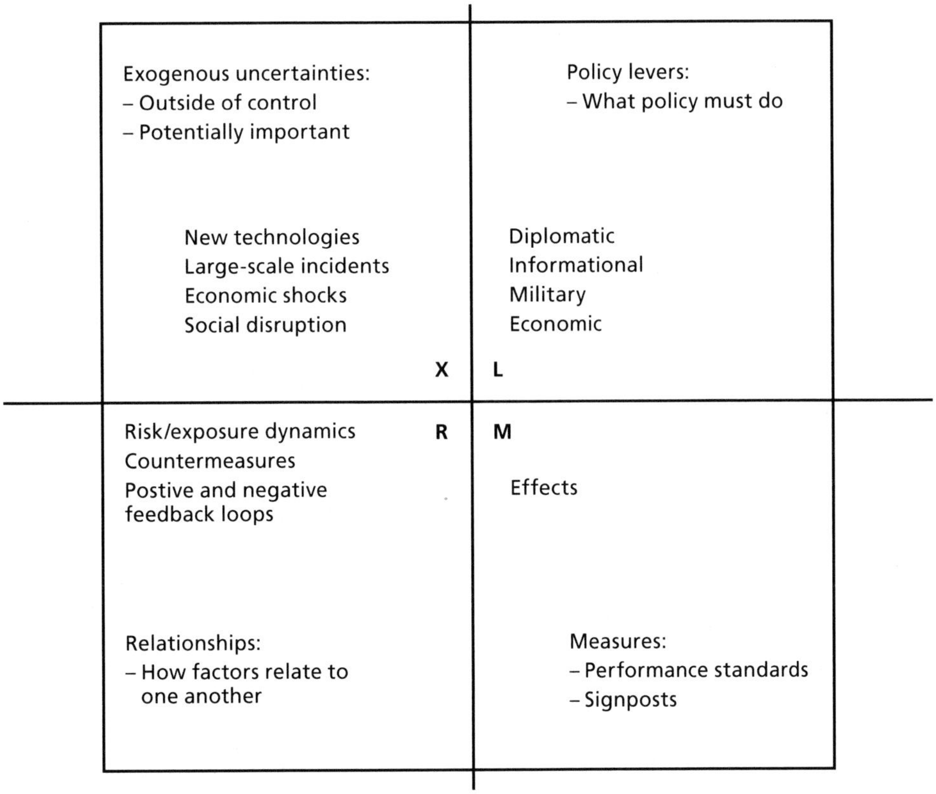

```
                    |
  +-----------------------+-----------------------+
  | Exogenous uncertainties: | Policy levers:       |
  | – Outside of control     | – What policy must do|
  | – Potentially important  |                      |
  |                          |                      |
  |                          |                      |
  |    New technologies      | Diplomatic           |
  |    Large-scale incidents | Informational        |
  |    Economic shocks       | Military             |
  |    Social disruption     | Economic             |
  |                          |                      |
  |                     X    | L                    |
  +--------------------------+----------------------+
  | Risk/exposure dynamics R | M                    |
  | Countermeasures          |                      |
  | Postive and negative     |    Effects           |
  | feedback loops           |                      |
  |                          |                      |
  |                          |                      |
  | Relationships:           | Measures:            |
  | – How factors relate to  | – Performance standards |
  |   one another            | – Signposts          |
  |                          |                      |
  +--------------------------+----------------------+
                    |
```

RAND *TR422-5.5*

The exogenous factors of Figure 5.5 (top left) include both certain and uncertain factors that are germane to the decision at hand. Policy levers (top right) are the options that are open to the various actors, both friendly and enemy. The commander's use of levers is therefore incorporated in the candidate courses of action; evaluation of those courses of action will consider the adversary's potential use of levers. Relationships (bottom left) are the ways in which exogenous factors are connected with each other and the ways in which policy levers affect the world; they reflect "theories" of effects. Measures (bottom right) are ways of assessing the world to ascertain, as clearly as possible, how desirable (or undesirable) any outcome is from the point of view of any of the actors.

This brings us back to effects-based operations, which emphasizes developing operations to achieve specified goals. In this framework, the measurement answers the question of how one will know whether the intended effects have been achieved or, before the issue is resolved, whether intermediate stages are in the right direction. Measurement is crucial for enabling the decision cycle to be restarted if the observations are too far distant from the intentions.

Scenarios for Foresight

In the absence of an accurate vision of the future, planners can construct one or more suggestive visions, often called *scenarios*, and then plan on the basis of them. There are many ways of defining and using scenarios, and a review of this burgeoning field (van de Riet, van het Loo, and Kahan, 2005) would take us beyond the scope of this report. Instead, we describe one way of defining scenarios for use in qualitative foresight exercises, one that has proved useful in a number of domains and that has potentially useful applications for command decisionmaking.

We define a scenario as a logical and consistent picture of the future. It is concrete in the sense that it has a level of detail that will make it amenable to quantitative or qualitative analyses. It can represent an extrapolation from the present based upon trend analysis or a discontinuity from the present and anticipated trends. It must be logical, internally consistent, and at least plausible.[3] A precise detailed scenario may be deemed very unlikely because of the concatenation of elements that are individually not very likely, but it should nonetheless be plausible (i.e., not impossible).

The central tenet of foresight work is that there is no such thing as a single best scenario. Because it would involve the concatenation of many elements, even the allegedly most likely scenario is a low-odds projection and a bad bet; therefore, multiple scenarios are the foundation for foresight analysis. The number needed may be very large, especially if the analyses are computer-based, using combinations of many factors, or it may be small if the analyses are largely qualitative, as described here.

A common tendency is to choose scenarios that are intrinsically interesting or that are somehow polar opposites; this is akin to the "orthogonal variables" approach critiqued in Chapter Three. Sometimes there is a tendency to "pick a future," i.e., to decide which of the interesting or opposite scenarios to use in planning. Our approach is different—we want the course of action to prepare us for anything that may happen within the "space" of scenarios that we consider plausible enough to be taken seriously. Because the space of possible scenarios is infinite, the challenge is to define a small number of scenarios that, if used to challenge our planning in different ways, will provide adequate insight into the larger scenario space of interest. No one scenario will be regarded as correct or even representative of the future; however, the set will be regarded as posing representative issues and challenges.

There is no clear-cut theory on how to identify the appropriate set of scenarios; instead, doing so combines theory with subject-area knowledge and art. We illustrate the approach later in this chapter.

The creation of scenarios can be described as a set of tasks: expanding, structuring, focusing, assessing, and constructing.

Expanding. The first step is to expand as much as possible the evidence base that will be used. Anything that could be a consideration in determining courses of action (L factors of

---

[3] We use *plausible* to mean *conceivable*, i.e., *possible*, as is customary in much of the strategic-planning literature. Something need not be *easy* to believe (i.e., credible) for it to be plausible. Regrettably, the word's basic dictionary meanings are different, such as "believable and appearing likely to be true" and "having a persuasive manner in speech or writing, often with an intention to deceive" (Encarta World English Dictionary).

XLRM) or relationships (R factors) is included; at this point, it is better to capture something that will later be abandoned than to miss something that would have been useful. Put another way, during the expansion task, we attempt to identify anything that could possibly make a difference in choosing among courses of action. This step involves divergent thinking, i.e., thinking away from the norm, in order to anticipate as much as possible.

Structuring. The next step is to put form onto the information that has been collected. Form consists of determining the major analytic dimensions that characterize the evidence, where the attributes of the dimension appear to be able to make a difference in choosing among courses of action. Examples of such dimensions include geographical and temporal scope, ability to influence the enemy course of action, capability of own troops, and political reactions to unfolding events. The methods used in structuring can range from multidimensional scaling techniques to content analysis.

Experience has shown that the expanding and structuring of tasks is not strictly linear in time but is interactive and can greatly be assisted by the construction of a relational database, especially one that exists well before the commander has to make a decision and that can be updated on a regular basis. At the present time, identifying items that should go into such a database and inputting them into the database are manual efforts, but with increasing technological sophistication, some of this can be automated. Moreover, the technology already exists to facilitate defining and using new dimensions within existing evidence. Currently, the relational database is in the collective mind of experts. For example, the mention of Iran may immediately trigger concern about the Strait of Hormuz and the oil supply.

Focusing. To maintain coherence and comprehension, the foresight exercise must have a focus. Focus is purposive—based upon objectives and not just on the inherent characteristics of the evidence. Focus is accomplished by introducing the effects that are desired and the commander's viable courses of action into the picture. For commanders who are open to multiple courses of action, all of the viable candidates are brought into the picture. For commanders who already have a specific course selected (perhaps only in broad outline), that one course is the point of focus.

Assessing. The assessment task is where the relationships among the uncertain dimensions of the scenarios are merged with the aspects of the future taken as certain and the courses of action to be assessed, in order to identify the number and nature of the scenarios to be employed. Once major analytic dimensions and the focus have been identified, the factors that drive success in achieving the desired effects need to be assessed as to the extent of uncertainty that exists and the importance of the dimensions. The database can be queried to ascertain what is and is not known for any dimension or combination of dimensions coded in it. This analysis and its purpose are shown in Table 5.1. From this analysis, we emerge with an idea of what must go into each scenario. Important features that are pretty much certain are put in the same form into each scenario. Unimportant features, whether certain or not, take a supporting role—provided to give concreteness and the impression of distinctness to the scenarios, or ignored if their use is inconvenient. The crucial cell in terms of determining which scenarios to build is the one that contains important and uncertain dimensions.

Table 5.1
Elements of a Scenario

Element	Important	Unimportant
Certain	Include in all of the scenarios	Use selectively to give a concrete picture
Uncertain	Identify key elements that differentiate scenarios	Mix and match to give color to the scenarios

Constructing. Finally, when the number and nature of scenarios are determined, the scenarios are constructed. While this statement is simple, the devil is in the details, which have been established in the previous four tasks. For the purposes of this report, there will be a small number of scenarios—two at a minimum and probably no more than four. Constructing these scenarios is as much an art as a science; it consists of preparing a concrete briefing specifying all of the details that would be presented if the scenario were ground truth. Each scenario should be concrete, logical, meaningful, and thought-provoking. If possible, and especially for highly uncertain important elements, some degree of drill-down should be built in anticipation of requests for more information. It should be emphasized that the construction of scenarios is not an end in itself, but only a means for managing uncertainty while choosing a specific course of action in the full detail needed for implementation.

Using Foresight Methods in the Command Post: An Illustrative Case

We present here an example of how a foresight approach could use scenarios to assist a commander in creating and evaluating courses of action. The specifics of the example should not be taken seriously, except as an illustration of the foresight concept. In this example, the commander wishes to obtain the effect of restoring order in the capital of a country being torn apart by civil war. The commander has at his disposal joint land and air NATO forces. None of these forces are present in the capital yet, but the area is surrounded and close to being isolated. Military force has been authorized by U.S. and NATO political leaders, but continuing authorization from all coalition partners is dependent upon not sustaining many casualties—either own forces or civilians. This particular commander is considering three possible courses of action and is not confident enough of any of them to express a preference yet. Instead, the commander realizes that success of the possible operations is dependent upon factors that have a great deal of uncertainty. The actions being considered are

1. *Direct attack.* A direct military action to occupy the capital and defeat the various fighting forces inside it.
2. *Isolation.* Fully isolating the capital so that the factions cannot be reinforced or resupplied, while engaging in an intensive information operation to influence the forces and civilians in the city by promising peace and reconstruction.
3. *Bribery.* Entering into an agreement with one of the warring factions to assist it in taking over the capital, in return for a promise of a democratic regime.

Because the crisis has been building up for months, a substantial database has been constructed. The evidence and analyses from the expanding and filtering tasks of the foresight analysis have identified several important aspects that are reasonably certain, as well as major uncertainties in the near-term future. The certain features include

- None of the factions possess chemical or biological weapons.
- None of the factions are strong enough by themselves to win a civil war; if there is no foreign intervention, the country will collapse into chaos.
- Coalitional lines of communication will remain secure in any operation.
- American political will is strong; a large majority of the people, including both political parties, are supportive of the American presence.
- Although civil war and collapse are imminent, if order is established, the country can rapidly achieve productivity and economic stability.

The uncertainties that seem initially to define the major dimensions of the scenario design are

- *Enemy sustainability.* It is not known how long the factions can continue the struggle without reinforcements and resupply. The best evidence points to time intervals of between three weeks and six months.
- *Coalition sustainability.* It is not known how long the NATO coalition can hold together. A close election in one of the major alliance countries scheduled for three months from now could cause that country's withdrawal and the possible dissolution of the coalition.
- *Faction trustworthiness.* It is not known whether the faction possibly amenable to assistance in return for moving ahead in a democratic fashion is capable of—or even desirous of—fulfilling its part of the bargain.
- *Enemy force strength.* It is not known how much resistance the forces in the capital can muster against occupation. It is possible that the factions might unite and together turn against the NATO forces.
- *Likely civilian behavior.* It is not known whether the capital's civilian population prefers outside occupation—even temporarily—to the factions fighting for control.

In the focusing and assessing tasks of the foresight analysis, it is determined that the faction-trustworthiness and enemy-force-strength dimensions are closely related—the more trustworthy the potential ally faction is, the less likely it is that the factions will unite. Additionally, the enemy-sustainability dimension is related to the coalition-sustainability dimension—the coalition does not appear to have the stomach for a long, protracted effort. Finally, the civilian dimension, in the analysis, takes on a dominant role, in that if the civilian population will welcome the coalition occupation, faction-trustworthiness uncertainty and enemy-force-strength uncertainty become of minor importance. This leads to the conclusion that civilian reaction and enemy sustainability are the principal factors. With further pondering, the team concludes that the following three scenarios would represent the scenario space reasonably well:

1. *Paris in the spring.* In this scenario, the civilian population would react to coalition take-over much as Paris welcomed the allied forces in 1944. The enemy force is not able to overcome this attitude, and even the factions most opposed to the coalition are likely to seek to get what they can get from the occupation. Under these circumstances, enemy sustainability is low and coalition sustainability is high.

2. *Porcupine love.* The name for this scenario comes from the old joke about how to make love to a porcupine—carefully. In this scenario, the civilian population will not automatically welcome the coalition but will be neutral to hostile to foreigners. However, there is good reason (not guaranteed) to believe that the potentially allying faction would be trustworthy, which means that any effort—either direct action or isolation—will likely be of short duration, and the coalition will remain intact.

3. *A policeman's lot.* The name for this scenario comes the observation that one of the most dangerous situations a police officer can be in is intervening in a marital fight. There is a significant likelihood that the couple will stop fighting each other and jointly attack the officer. In this scenario, there is evidence that the various factions might combine to attempt to drive the foreign forces out (and then, presumably, return to their internal squabbles). Moreover, the civilians are likely to be part of this united front. Things could get messy.

Figure 5.6 shows how the three scenarios are rated according to the five dimensions of uncertainty. Figure 5.7 recasts this information by portraying the extremes on each dimension,

Figure 5.6
The Three Scenarios in the Five-Dimensional Uncertainty Space

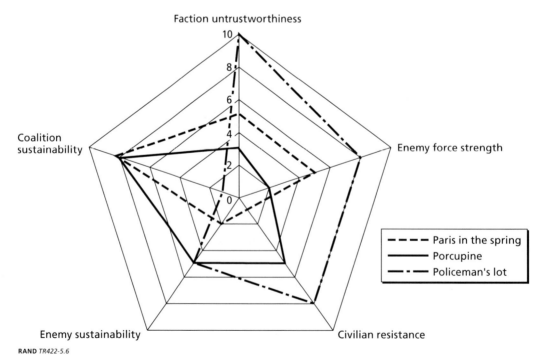

RAND *TR422-5.6*

Figure 5.7
The Space of Possible Futures to Consider

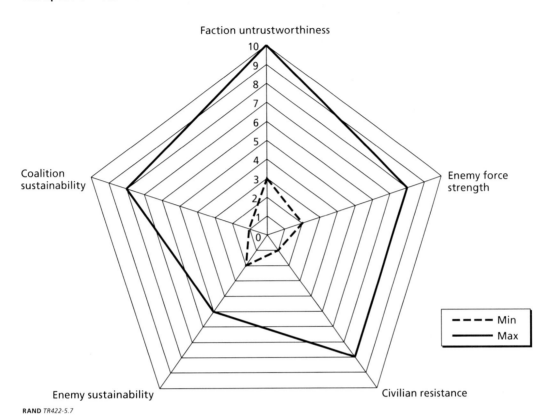

RAND *TR422-5.7*

regardless of which scenario generated it. The space defined in Figure 5.7 between the minimum value and maximum value for each dimension, over the five dimensions, represents the bounded area within which we assume the future will lie. Part of the possibility space has been excluded because it is regarded as too improbable to be considered, even in conservative planning.

Each of the scenarios is assigned to a writer, who develops it in the form of a command-post situational briefing. Elements that are constant in each scenario, including the items mentioned above plus potentially important endogenous factors such as own-force readiness and information-operations assets available for use, are also incorporated into the briefing.

The commander then divides his command staff, as equally as possible, into three teams. Each team is given one of the three briefings and is then given two hours to analyze the scenario and answer the following:

- *Question 1.* Given the desired effect of restoring order in the capital with as little collateral damage as possible, how well will the operations resulting from each of the proposed courses of action do, within the scenario considered?

- *Question 2.* To the extent that uncertainties remain in determining the answers to Question 1, what are these uncertainties and what evidence that could be gathered might reduce them? This exercise should be undertaken only if the resolution of the uncertainty might change the preferred course of action.
- *Question 3.* Are there hybrid or completely new courses of action that might be better for this scenario than the ones posed? If so, what are they?

Question 1 is the traditional one asked of a scenario analysis and requires no elaboration here. Question 2 is crucial, since any given scenario may not play out as expected (or as suggested in the scenario concept). What might go wrong, or right? This question encourages the teams not only to take into consideration the uncertainties previously presented, but also to consider additional ones that arise in their analysis of the effects of the courses of action. For example, in the porcupine scenario, it was assumed that the faction would be trustworthy. But in evaluating the courses of action, two considerations arise. First, if one faction is trustworthy, this may influence the others to team up against it. Second, the trustworthiness apparently depends on the maintenance in power of the faction leader; if he is assassinated, things could change drastically. This additional uncertainty is taken into account, so that the recommendations of the porcupine team (and the other teams as well) contain some hedging. Question 3 is also important, and takes the commander beyond the constraints of considering each course of action as an orthogonal alternative.

Finally, the commander meets with all three teams together and receives their scenario-dependent analyses as answers to the three questions. Although, in some sense, the recommended course of action for each scenario in this example is pretty much determined by the scenario (acknowledging that there could be some surprises), it is the deeper analysis that is of more interest. In essence, the teams have done their work from the phenomenology of each individual scenario. The focus now shifts to the phenomenology of uncertainty among the scenarios (as well as all potential situations that are in the space defined by the points of the scenarios). The analysis in this session proceeds as follows:

1. Is there one course of action that is best for all of the scenarios? This is, of course, unlikely, but it should still be asked. If so, then the exercise has been completed.
2. How do the strengths, weaknesses, opportunities, and threats differ from scenario to scenario, and how do they remain the same? Aspects that remain the same may be considered robust and an important part of the decisionmaking process. For those aspects that differ, what are the scenario dimensions behind the differences?
3. Before fully settling on a course of action, are there intermediate steps that can be taken to turn the situation toward some of the desirable features of some of the scenarios? That is, if one of the scenarios has more strengths and opportunities (or fewer weaknesses and threats) than the others, could matters be aimed toward that scenario?
4. Are the revealed additional uncertainties the same or different by scenario? Those that are the same yield high priorities for further information-gathering efforts. How many of these additional uncertainties might be resolved before a course of action *must* be

chosen? What are the risks of going ahead with a course of action with these uncertainties unresolved? To what extent can a course of action be altered as new information emerges?

5. Consider constructing hybrid or alternative courses of action that are robust (that is, have an acceptable chance of achieving the desired effects) under multiple scenarios. In other words, is there some form of a course of action, possibly incorporating preliminary steps and further information, that might be acceptable under all scenarios?

6. Given all of this, what appears to be the best course of action? What risks does it entail? How adaptable is it to unfolding circumstances and new information?

At this point, the meeting reverts back to a more familiar decisionmaking process, in which the chosen course of action (either one of the three original courses, a hybrid, or something completely different) is fleshed out in terms of how it can be implemented. If the additional uncertainties are so large that the original analysis does not lead to a firm conclusion, a second assessment is undertaken, following the outlines of the portfolio analysis described above, to further zoom in on the uncertainties.

Table 5.2 illustrates how this iterative assessment might be worked out, based on the initial courses of action and on variants modified either to improve prospects for best-case outcomes or to moderate the worst-case outcomes. For example, the direct assault might be modified to provide excellent, persistent air cover for the ground forces, as well as preplanned exit procedures in the event of greater-than-anticipated trouble on the ground, as might happen in either the porcupine or policeman scenario. The scores are purely hypothetical, but if the assessments came out as shown, the modified version of the second option, isolation, would

Table 5.2
Evaluation of Courses of Action, Before and After Iteration in a Foresight Exercise

Course of Action	Paris in the Spring			Porcupine Love			Policeman's Lot		
	Worst	Most Likely	Best	Worst	Most Likely	Best	Worst	Most Likely	Best
Direct attack	4	8	8	3	5	8	1	4	5
Isolate	4	7	7	3	7	8	4	5	6
Bribe	5	6	8	4	5	6	4	6	7
Direct attack	4	9	10	4	5	10	3	4	6
Isolate	4	7	10	3	7	9	4	5	6
Bribe	5	6	8	4	5	6	4	6	7
Direct attack	Averages of most-likely outcomes across scenarios								M 5.4
Isolate									G 6.3
Bribe									M 5.7
Direct attack	Averages of best-case outcomes across scenarios								VG 8.7
Isolate									VG 8.3
Bribe									G 7

NOTE: VB, B, M, G, VG = very bad, bad, marginal, good, very good.

be the winner. Although the modified version of direct assault would have a somewhat higher upside potential, it would not be enough to compensate for isolation's advantage in the most likely versions of the scenarios. All of this, however, assumes that the commander chose to treat the scenarios as equally likely. If he completely discounted the worst cases and took the upside cases quite seriously, then the modified version of direct attack would be best.

CHAPTER SIX

Priorities for Investment in Families of Models, Games, and Other Tools

Introduction

Chapter Five discussed two particular methods/tools in some detail. In this chapter, we discuss more generally the kinds of methods and tools that we believe are needed for decision-support systems serving operational commanders and defense planners.

It follows from the demands described in Chapters One through Four that a number of key *functional needs* can be identified for analysis and supporting modeling and simulation. These include (1) routine and perceptive treatment of uncertainty, (2) emphasis on FAR strategies, (3) adaptive models, and (4) reinserting people in modeling and simulation and related analysis.[1] The needs are not merely nice-to-haves; rather, they lie at the core of whether decisionmakers are to be well served. This chapter deals with some of these under the headings of adopting a more holistic family of tools, building low-resolution models for exploratory analysis, and improving model adaptiveness.

Families of Tools

The Concept of a Family

An appropriate family of tools to support analysis would include (whether in one location or in the virtual world) at least the following, each of which has its own advantages and disadvantages:

1. A diversity of models with different levels of resolution, perspective, and character and different degrees of interactivity.
2. Human games and other exercises structured to increase rigor and analytical content.
3. "Laboratory" and field experiments; and

---

[1] This material draws on work done in a project for OSD's Office of Program Analysis and Evaluation (PA&E) as part of OSD's effort to develop a master plan for the analysis component of modeling and simulation (Davis and Henninger, 2007).

4. Other empirical work, drawing upon real-world operations planning, training, history (including lessons-learned studies), and consultation with experts in numerous disciplines and functions.

Figure 6.1 elaborates by contrasting the instruments' strengths. At the top left, analytical models[2] are characterized as very good (white) for their agility and breadth, i.e., their ability to generate a synoptic view and respond quickly to a myriad of "what-if" questions. They are also reasonably good (very light gray) for decision support, for the same reasons. They fall short (black) in other respects. As a contrast, field experiments are good or very good for integration and representing phenomenology, including human issues (bottom right) but are poor for agility, breadth, and decision support.

Figure 6.1
Strengths and Weaknesses of Different Tools in the Family

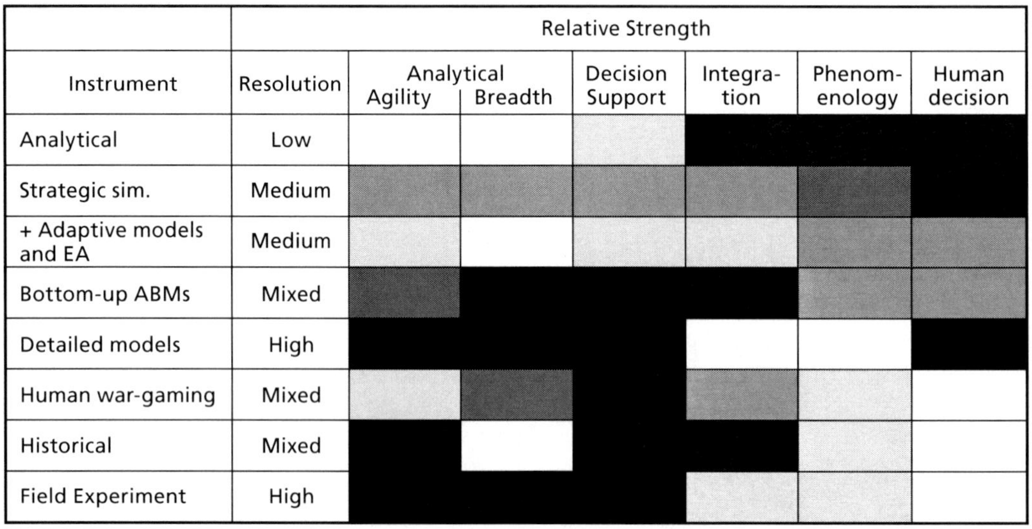

Instrument	Resolution	Relative Strength					
		Analytical Agility	Breadth	Decision Support	Integra-tion	Phenom-enology	Human decision
Analytical	Low						
Strategic sim.	Medium						
+ Adaptive models and EA	Medium						
Bottom-up ABMs	Mixed						
Detailed models	High						
Human war-gaming	Mixed						
Historical	Mixed						
Field Experiment	High						

Coding: Black, dark gray, gray, light gray, and white are very poor, poor, marginal, good, and very good, respectively. Colors depend sensitively on many implicit assumptions. Readers should focus simply on the point that instruments have different virtues.

ABMs: agent-based models
EA: exploratory analysis

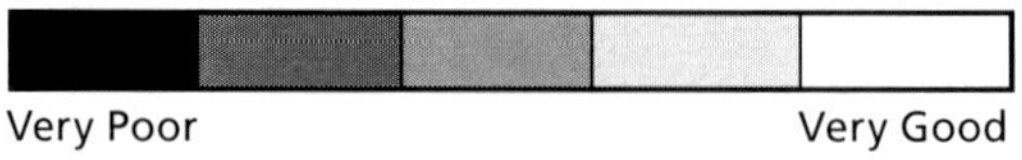

Very Poor   Very Good

NOTE: Figure adapted from Davis (2005).
RAND TR422-6.1

---

[2] In this context, *analytical models* refers to relatively simple and transparent models such as might be implemented in an Excel spreadsheet (Ochmanek, Harshberger, Thaler, and Kent, 1998) or the Analytica system. They usually provide highly aggregated descriptions of phenomena, although in some cases, they can instead describe the essence of a detailed physics problem well, such as understanding the dependence of long-range precision fires on the nature of terrain, the target's maneuver pattern, timeliness of command and control, and weapon characteristics (Davis, Bigelow, and McEver, 2001). Such models may have from five to 20 key parameters, but not hundreds or thousands, plus complex databases of allegedly fixed data as are more typical of DoD modeling and simulation.

Strategic simulations (e.g., the Joint Integrated Contingency Model (JICM) and Thunder) are concerned with theater and multitheater analysis, primarily at the theater-strategic and operational levels. In principle, these can have moderately good capability for analytical functions, decision support, and integration. In addition to other flaws, however, current versions of such models are not very adaptive and are still typically used—despite guidance to the contrary from OSD—to study particular approved scenarios in depth, thereby sweeping large uncertainties under the carpet. If the models were more adaptive (as discussed later), were improved in other respects, and were used for broad exploratory analysis, their value could rise a notch to light gray or even white.

Agent-based models of the bottom-up variety have at least moderate ability to explore phenomenology and human action, although analysis with them is not straightforward (Sanchez and Lucas, 2000). In other respects, they are typically quite limited.

Detailed models (e.g., those at mission, tactical, or engagement levels) are essential underpinnings even for higher-level defense-planning work, because they can represent underlying phenomena well, motivate sound aggregate models (e.g., strategic simulations), and sometimes be used to calibrate them. Such models are also an essential link between the "policy and programs world" and the world of war-fighters and engineers. Such models, however, are poorly suited to higher-level analysis or decision support. Further, despite their detail, they are as subject to uncertainty as the higher-level models. They can be reasonably predictive, but only for highly circumscribed assumptions.

Human war-gaming is quite agile in the sense that a game can be quickly put together to deal with previously unstudied issues, but games have been notoriously deficient in breadth when they have employed only a single scenario or a few vignettes. Multiple scenarios, as described in the previous chapter, can improve the situation to a small extent. Games are, however, excellent vehicles for highlighting "real" factors in the world, including likely or possible human perceptions and behaviors. A human game, for example, will often produce an unanticipated adversary thought pattern or an unanticipated constraint on what U.S. and allied leaders would be able and willing to sanction.

Historical work is not at all agile, but it covers substantial ground. It is not good for decision support or integration, but it is a rich source of knowledge about what "really" happens, what humans sometimes do, and how frequently they make mistakes in some cases but achieve audacious success in others (Keegan, 1983; Gordon and Trainor, 2006). Some historical research provides empirical evidence of aggregate-level phenomena (Dupuy, 1987), which can be powerful in developing models and simulations. Crucial to any historical analysis is the connecting of the variables of what was then to the variables of what is today (Neustadt and May, 1986).

Field experiments are often considered the gold standard, and they can indeed be excellent (white) for integration, seeing how operations would really unfold, and revealing human issues; however, this comes at a considerable cost to agility and breadth, because few variations can be employed in any experiment.[3]

Although the evaluations shown are merely illustrative, depending on a great many underlying assumptions, the general point is valid: There is much value in drawing upon multiple instruments and sources of information. It follows that analytical organizations should consciously plan to have and exploit the full range of instruments and forms of inquiry.

One conclusion from thinking about the family-of-tools approach is particularly important:

- Because humans are so much better than current modeling and simulation at many crucial functions involving, e.g., creativity, finding "holes" to exploit in plans or weaknesses in an adversary's posture, first-order thinking through of operational concepts, and cross-cutting work, it is essential to increase the role of humans in analytical activities for decision support.

This conclusion is only strengthened by today's emphasis on operations in which PMESII factors and DIME instruments loom large.[4]

Implementing this recommendation will not be easy, in part because a generation or two of analysts and decision-support systems have emphasized dependence on "closed" computer models with no human interaction except by analysts establishing inputs and examining outputs.

Because we consider the human role to be so important, we presented in Chapter Five a new example of creative gaming (the *foresight exercise*) adapted from experiences in social science, one that we believe could be used by operational commanders (although not on the eve of battle). We next address a number of other ways in which humans could be better used .

Making Better Use of People in Connection with Modeling and Simulation, Analysis, and Decision-Support Systems

How can humans be used more effectively in modeling and simulation, analysis, and decision-support systems? Many ways come to mind as soon as one thinks about the matter. These include, e.g., various types of human gaming, use of experts and other thoughtful people in building models, optionally interactive war-gaming, and drawing upon expert panels for special issues. We address each of these in turn in the following subsections.

---

[3] Although rated highly in Figure 6.1 because some are indeed so good, most field experiments are much less effective for the purposes indicated in the figure. Their real value is often in convincing leadership (and rank-and-file officers and crew, as part of training) that certain capabilities now exist, with demonstrations rather than experiments, or in influencing allies, potential adversaries, and other observers. These are also very worthy purposes, but they are not particularly relevant to this report. Reviews of both joint and service-level experiments often recommend more effort on smaller-scale exercises and increased use of models, simulations, and analysis (National Research Council, 2004; Defense Science Board, 2003a).

[4] This point was a prominent theme in a recent National Academy study of DoD modeling and simulation (National Research Council, 2006).

Human War Games, Including "Analytical" Versions. A natural question is whether human gaming can in practice be used to inform analysis. The stereotype is that human games are idiosyncratic to players, focused on the playing through of a single scenario, undocumented, and relatively unstructured. In fact, however, "gaming" refers to a broad range of very different kinds of activities, so generalizations are unwise. Human gaming can be used for at least the following analytical purposes: discovery, sensitization, concept development, knowledge elicitation, identification of assumptions, and testing of hypotheses.

Examples are many, if one merely looks for them. Armies have long used war-gaming to test hypotheses and strategies and sometimes to discover new ones. James Dunnigan and others drew heavily upon study of military history to create recreational war games that were used by young officers well before today's era in which analogs are computerized (Dunnigan, 2003). Aspects of the JICM model (e.g., soft factors and a network treatment of maneuver) were motivated by the strengths of human gaming (Allen and Wilson, 1987). RAND's "Day After . . ." games (Molander, Wilson, Mussington, and Mesic, 1998) have been used for 15 years for initial cuts at serious policy problems, such as nuclear use and cyberwar. The U.S. Joint Forces Command routinely uses a model-game-model approach. And quite recently, OSD (PA&E) has used gaming to study issues in the resourcing for and employment of special operations forces. In the academic domain, Michael Zyda at the University of Southern California, previously the director of the Modeling, Virtual Environments and Simulation (MOVES) Institute at the Naval Postgraduate School, has sketched the nature of a discipline that he calls "serious games" theory. Zyda defines serious games as mental contests, according to certain rules, played with a computer, that use entertainment to further government or corporate training, education, health, public policy, and strategic-communication objectives.

How to make human war-gaming more "analytic" deserves an entire paper, but we did present one example (the *foresight exercise*) in Chapter Five. Another approach has been sketched as follows (see Figure 6.2 below) (Davis, 2004):[5]

- Design the games as vignettes with relatively well-described situations.
- Use competing teams with different backgrounds (e.g., from the United States, the UK, Israel, and Poland) to see diverse tactics and assumptions.
- Encourage teams to develop explicitly contingent plans (e.g., with branches and sequels).
- Protect the teams organizationally, perhaps by embedding them in independent groups such as federally funded research and development centers (FFRDCs), war colleges, or the U.S. Joint Forces Command.
- Record planning factors and reasoning used during team play, recognizing them as the germs of "models" that can later be formalized, either quantitatively or qualitatively (Davis, 2001; Davis, 2002b).

---

[5] The astute reader will notice that the foresight exercise described in Chapter Five has many, but not all, of the attributes suggested here.

Figure 6.2
A Process for Using Human War-Gaming Analytically

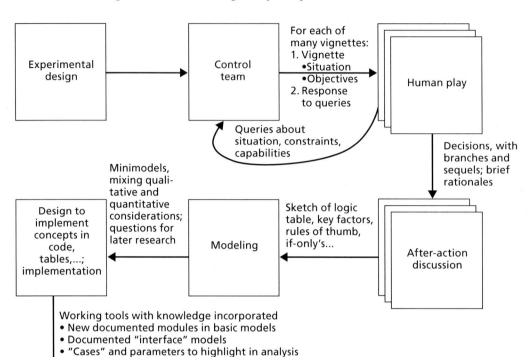

RAND *TR422-6.2*

- Use Red teams, both to better appreciate different ways of assessing the situation and defining objectives and to draw upon expertise about adversary military doctrine.[6]
- Follow up with analysis and modeling.

Analysis should occur at the beginning and the end of Figure 6.2. People skilled in capabilities analysis should design the war games and vignettes to cover the space adequately (perhaps with a combination of experimental design and of modeling and simulation greatly narrowing the number of cases). Analytically inclined people with an appreciation of and openness toward soft factors should record intragame logic and then develop summary qualitative models. Subsequently, modelers should relate the variables of the human games to variables of models and simulations.

---

[6] Red-teaming has been reviewed in some recent efforts by the Defense Science Board (Defense Science Board, 2003b); a Hicks and Associates study of counterterrorism led by James Miller (Murray, 2002, 2003; Sinnreich, 2002; Murdock, 2003); and unpublished RAND work by the senior author.

Such an agenda does not require any breakthroughs in science or technology. It merely posits that good traditional quantitative analysts bring their skills to bear on a problem that has often been studied in the past by historians, political scientists, and military officers without the benefit of much formal analytical training.[7]

Using Human Gaming to Help Inform the Building of Adversary Models. A different approach for using humans and gaming to help inform and tune adversary models starts from a theoretical structure (Davis, 2002b). In this approach, gaming is used to test an initial model and suggest additional variables. Procedurally, the approach is as follows:

- Develop a theory and structure for understanding possible high-level adversary decisions and behaviors.
- Use political-military seminar war-gaming to check on the adequacy of the factors and structure and to test the theory in particularly difficult or ambiguous situations.
- Iterate the theory and use it to generate alternative adversary models, each of which is parameterized to reflect inherent uncertainties, even for specified conditions.
- Use these adversary models with exploratory analysis to develop or test candidate FAR strategies.

Precedents for this approach exist at several levels of detail. The same methods can also be used to build models of U.S. or third-party decisionmaking.

Another very encouraging example of how adversary modeling can be accomplished focuses on inferring the intent and developing the consequences of that intent for subsequent actions in a dynamic environment (Santos and Zhao, 2006). This approach has been used in military war-gaming.

Using People in Optionally Interactive Simulation in Building War Plans. A third approach is optionally interactive simulation, where humans may (or may not) be used to make command-and-control decisions, such as shifts in strategy or commitment of reserves. At one time, such human intervention was necessary, but as modeling and simulation improved, the aspiration was to eliminate human play so as to improve reproducibility and perceived rigor. In some systems (including the original Joint Warfare System (JWARS)), interactivity was explicitly proscribed; in others (so-called interruptible simulations), it is technically possible at some natural points but seldom exploited well. The tack suggested here is as follows:

- Have humans sketch out the "war plans" to be used in simulations.
- Represent the plans in the simulation.
- Conduct the simulation with interruption points at which humans review the situation and, as necessary, make adjustments in the strategy by identifying what amount to new building-block actions or triggers for actions.

---

[7] The Army War College has long developed databases on the forces that have historically been used in various contingencies. A recent RAND study (Dobbins et al., 2003) reviews the perceived historical need for forces in the stabilization and reconstruction phases. Such studies sometimes fail to convince skeptics, because they show what *was* used, rather than what was necessarily required or what may be needed today, but the information is nonetheless useful and often sobering.

- Add the new action sets or triggering rules to the simulation.
- Iterate until automated play represents well the strategy and adaptations of the humans.
- Repeat this process with different human teams so as to develop a set of alternative strategies and building-block actions for both Red and Blue and, as necessary, for third parties.

This mode of operation has some associated technical requirements. First, interruptible play must be possible. Second, strategies must be conveniently representable, to include branches and sequels triggered by state-dependent rules. Third, it must be possible to construct easily accessed libraries of building-block actions, a number of which might be used in any given strategy, perhaps conditionally. In past work, this style of man-machine operation was found to be quite effective.[8] It appears not to have been much used in recent years, however.[9] That should change.

Other Ways to Make Better Use of Experts in Modeling and Simulation. Another aspect of a holistic family-of-tools approach is making better and more systematic use of experts. Doing so overlaps with gaming, construed broadly.

There is no dearth of methods for using experts. Such methods include Delphi (Helmer-Hirschberg, 1967; Linstone and Turoff, 2002), the Analytic Hierarchy Process (Saaty, 1999), Value-Focused Thinking (Parnell, 2004), Subjective Transfer Function techniques (Veit, Callero, and Rose, 1984), Scenario-Based Planning (Schwartz, 1995), Day After Games (Mussington, 2003), Uncertainty-Sensitive Planning (Davis, 2003), and Assumption-Based Planning (Dewar, 2003).

Nor is there a shortage of mechanisms for representing expert knowledge in models. Examples include agents, influence nets (Wagenhals, Shin, and Levis, 2001), rules and sub-models in system dynamics (Simulex, 2006), and the plans represented in simulations.

Typical attempts to elicit expert knowledge have been less than satisfactory, however. As discussed at a recent workshop sponsored by OSD (PA&E), the principal difficulties have included the following (Davis and Henninger, 2007):

1. Finding the experts in the first place (if they even exist and are available for DoD work).
2. Recognizing and dealing with the fact that experts often have "agendas."

---

[8] Such operations were routine in the RAND Strategy Assessment System (RSAS), the predecessor of the current-day JICM model. The RSAS had libraries of "analytic war plans" composed of building-block operations (Schwabe and Wilson, 1990). It exploited a high-level language (RAND-Abel) and special structures motivated by artificial-intelligence research. JICM does not have these, but it has an interface language that provides much of the same functionality (Jones, 1999). The more-detailed JWARS system, now being tested by U.S. Joint Forces Command and others, has significant capabilities for user-defined rules that exist in data and are easily changed (Burdick et al., 2002). These are compelling existence proofs, whether or not they are complete and satisfactory as they currently stand.

[9] In contrast, the process of model-game-model has been used often and effectively over the years, dating back at least to the predecessor of J-8 (SAGA) and into recent years. For many years, the most careful Joint Staff analyses were done with TACWAR used as a game board and bookkeeper, but with military officers making key decisions along the way on force employment and sometimes on adjudication of results. Current work by the Joint Forces Command also uses a model-game-model approach.

3. Unfortunate group dynamics, such as effects of hierarchy and social context and the well-known group-think phenomenon.

4. Inappropriate selection, e.g., merely inviting the "usual suspects" or, worse, inviting only those expected to support a concept being evaluated.

5. Having effective cross-disciplinary discussions when experts from diverse disciplines often have different languages, assumptions, and tacit knowledge.

6. The insidious tendency in many contexts for the organizers of expert discussion, and perhaps the group of experts itself, to move toward a best estimate or consensus, rather than exploiting the opportunity to see distributions of possibilities or exploiting the so-called "wisdom of crowds."

Doing better is not necessarily straightforward, and there are contrasting paradigms about how to use experts. Some think in terms of the wisdom of crowds, while others seek the top experts; some favor the "blink" theory and others the "think" theory; and so on. The conflicts are sometimes more apparent than real, with different methods being suitable in different circumstances. Nonetheless, there is need to think about the issues seriously. The way ahead in this area should include the actions suggested in Table 6.1.

Let us now move away from the issue of using humans and address another aspect of achieving good families of models, the special challenge of obtaining the good analytical models suggested in the upper left of Figure 6.1.

Building Low-Resolution Models for Exploratory Analysis

A continuing bugaboo in efforts to develop families of models is that people and organizations accustomed to working with large, complex models often have no good ideas about how to obtain good *simple* models of the type preferred for broad-ranging exploratory analysis,

Table 6.1
Action Items for Improving Use of Experts in Support of Modeling, Simulation, and Analysis

Science and Technology	
Write the book	Develop a primer on how the various techniques for using experts apply in DoD work. Spin off learning materials for continuing professional education of both military officers and DoD civilians.
Specific issues	Find or develop better techniques for abstracting expert assessments from groups, without omitting significant outliers or seeking consensus but nonetheless eliminating counterproductive "noise."
Technology	Review technology available to best use consultants for DoD modeling, simulation, and analysis purposes, collocated or distributed.
Within Analytic Studies	
Specific studies	Include in terms of reference for major DoD analytic studies that experts will be used and, where appropriate, specify the kinds of expert advice being sought (e.g., specify use of independent Red teams and mechanisms to assure that their concerns are both captured and assessed).

the type that is so valuable in decision-support systems. Table 6.2 lays out four methods for doing so, the last of which is not to our liking because it tends not to illuminate the underlying phenomenology or "tell as a story."[10]

The first method is attractive if one has a model designed with multiple levels of resolution in mind. In these models, one can turn off higher levels of detail for particular modules when doing exploration and then turn them back on as needed to follow up on conclusions in more detail. Such features have been built into only some of the commonly available models. Adding them to pre-existing models, however, is straightforward in some cases.

Table 6.2
Ways to Create Simplified Models for Exploratory Analysis

Approach	Description	Comments
Freeze many variables of more-complex models; conduct exploratory analysis on a set of key parameters[a]	• Find or create higher-level parameters • Assure legitimacy of freezing other variables	Feasible only in some models, e.g., those designed top-down with multiresolution modeling techniques; for other models, important uncertainties are often too deeply buried in databases
Build a new simple model[b]	• Approach the problem top-down • Build in specific hooks to higher-resolution models • Parameterize extensively • Allow for structural variations as well	Building simple models is easy; building good ones, which abstract properly and connect to more-detailed models, is not
Build a motivated metamodel[c]	• Start with a "trusted complex model" • Hypothesize a much simpler model that is still phenomenologically motivated, but in aggregate terms; build in correction factors with unknown coefficients • Use statistical methods to fit the postulated model to the behavior of the trusted model	The "trusted model" may not include important dimensions of uncertainty, such as adaptive decisions; if trust is justified, however, this approach can lead to reliable simple-and-fast models that also provide good explanations of results
Standard response surface methods[d]	• Fit the behavior of the trusted model to statistical regressions	The resulting "simple model" provides no meaningful explanation and may obscure important correlations relating to system capabilities for systems with multiple individually critical components

[a] Much work of this type has been done with the JICM model (Fox, 2003).

[b] There are many examples throughout the community (Davis, Bigelow, and McEver, 2000; Davis, McEver, and Wilson, 2002), although most were one-time actions. This approach is also common in the physical sciences.

[c] This is a synthesis of methods from cause-effect modeling and statistical analysis (Davis and Bigelow, 2003).

[d] This approach is commonly taught in operations research and statistical methods courses.

---

[10] Some of these points were made in a recent study for the Navy's OP-81 (National Research Council, 2005) and in a full-day workshop on multiresolution modeling and exploratory analysis sponsored by the Military Operations Research Society (Davis, 2005b).

Improving Model Adaptiveness: Agents and Other Methods

General Approaches

As discussed earlier, it is essential that modeling and simulation become substantially more adaptive if they are to be able to inform the choice of strategies intended to be flexible, adaptive, and robust. The adaptiveness may be achieved by (1) having submodels representing decisionmaking by commanders, submodels that adjust simulated strategy and tactics depending on objectives, situation, and projections; or (2) having submodels representing, e.g., the behavior of individuals (perhaps adversary leaders), groups (e.g., a particular regional tribe or the military establishment of a dictator in a losing war), or countries (e.g., a country that might or might not become more nationalistic and resistant as the result of strategic bombing). These are only some of many examples.

The methods available for improving adaptiveness include using agents, control theory, game-theoretic methods, or more ordinary model-related operations-research algorithms. The methods may be deterministic, stochastic, or hybrids. These methods should be seen as being in competition. Unfortunately, the book has not yet been written on when each type of method is suitable.

Agent-Based Models

Agent-based models are currently of particular interest, in part because they are relatively new and in part because they show so much promise for dealing with certain kinds of issues, both tactical and in the domain of social-cultural-military interactions.[11] Many different types of agent-based models already exist, and other types are needed to fill particular gaps.

One important distinguishing characteristic relates to whether the agents are conceived top-down, as part of command and control, or bottom-up, as part of atomic entities that act independently but generate higher-level emergent behaviors (e.g., riots arising from the action of individuals in a crowd, groupings of tribes into coalitions, or steady-state mass behaviors of traffic flow). A second characteristic is whether the agents' internal structures are simple ("light") or sophisticated ("heavy"). A third characteristic is whether the agents are part of a simple larger simulation or a sophisticated one. Still other characteristics relate to programming style and language. The list goes on.

The principal observation here is that generalizations are dangerous in a domain as rich and with as much potential as agent-based modeling. Further, despite active and fruitful research over the past decade or so, the field is still young on the time scale of scientific developments.

This said, experience with bottom-up agents has been mixed. First, they are not easily generalizable, because they—especially those built bottom-up—embed much of their knowledge in deeply buried rules that are dependent on old contexts and must be reworked or new contexts such as different countries or cultural settings. A second problem is that cause-effect relationships of observed macro behavior are difficult to identify, because they are the result

---

[11] The recent agent literature is extensive and includes reviews (Uhrmacher and Swartout, 2003), applications to military problems (Ilachinski, 2004), and applications to social-network problems (Prieutula, Carley, and Gasser, 1998), among others. Project Albert, sponsored by the Marine Corps, has been a prominent effort (Horne and Leonardi, 2001). A broader view of agents is described by Davis (2005a).

of many micro-level relationships and events, which often occur in complex adaptive systems. A third problem is that the models typically do not have explanation capabilities, so they are used as black boxes or, to be more pejorative, as oracles. Finally, validation of such models is both difficult and unstable because of the many underlying rules and networks and because agent-based models are inherently inclined to generate behaviors that may be plausible and important but are not intuitive. This is a strength for open-minded research, but it undercuts the method of depending on face validity of behaviors for evaluation. The way ahead for agent-based models is suggested in Table 6.3 (Davis, 2005a; Davis and Henninger, 2007).

Table 6.3
Action Items for Use of Agent-Based Models

Science and Technology	
Invest with a portfolio approach	Invest in different approaches to agent-based modeling work (e.g., top-down, bottom-up, hybrids), in abstraction methods for representing agent-based-model behaviors in more analytical models, and in related tools
Write the book	Support writing of DoD-oriented review papers and books relating different agent-based modeling methods to different problems and functions
Validation	Define and promulgate new concepts of "validation," such as validation for the purposes of exploration
Practice	Develop a "code of best practices" for use of agent-based models in analysis
Applications	
Invest in diverse types of application	Recognize the virtues of applications for discovery, hypothesis generation, . . . , rigorous analysis
Enrich classic force-on-force models	Add decision models to provide adaptivity, probably using top-down agents
Enrich treatment of command and control in modeling and simulation	Use agent-based models as necessary to assure that modeling and simulation can represent core phenomena such as shared awareness and self synchronization
Supplement combat-phase models with models of, e.g., stabilization and reconstruction	Use bottom-up agent-based models (or possible abstractions) for modeling of phases in which emergent behaviors are particularly important
Broaden participation	Assure participation of, e.g., Department of State, Department of Commerce, and other organizations, or of relevant experts
Encourage competition and openness	Invest in both open and classified applications, encouraging as much openness and peer review as possible

CHAPTER SEVEN

Conclusions

In this report, we have described the rationale behind an approach to high-level decision support, such as is needed by a Joint Forces Air Component Commander (JFACC) in operations or a Chief of Staff in defense planning. Our central theme has been the fundamental importance of dealing effectively with uncertainty, whether in effects-based operations, building the Commander's Predictive Environment, or planning future forces with capabilities-based-planning methods.

We have described what we see as general principles for a decision-support system serving high-level military decisionmakers, and we have proposed a way in which they can be applied even by decisionmakers with very different personal and cognitive styles. We have gone on to illustrate the principles with more-concrete examples, one derived from portfolio-management methods and one adapted from a form of human gaming that provides more structure and deeper insight than is usual in such activities.

Finally, we have drawn implications for investments in basic research and technology enabling decision-support systems and the use of modeling and simulation. A high priority should be placed on methods for discovering, constructing, and evaluating flexible, adaptive, and robust (FAR) strategies. This leads to requirements for adaptive models, which can be achieved by drawing not only on the new field of agent-based modeling, but also on control theory, game-theoretic methods, and traditional operations research. In order to achieve FAR strategies, however, it will often be necessary to exploit the creative, cross-boundary abilities of people. We believe that it is essential that decision-support systems be designed to use not only traditional modeling and simulation, but also human-intensive methods such as war-gaming, foresight exercises, Red-teaming, assumption-based planning, and various methods for using experts well—not for prediction, but rather to identify possibilities and ways to prepare for them.

Appendix

Issues and Controversies Regarding Effects-Based Operations and the Commander's Predictive Environment

Background

Despite the steady movement to incorporate effects-based thinking in both Service and joint doctrine, and despite the recent emergence of pre-doctrinal materials (U.S. Joint Forces Command, 2006a,b), effects-based operations (EBO) has recently suffered a setback, with the Commander of Joint Forces Command (JFCOM) backing away from it temporarily, or at least taking a fresh look at it in the wake of criticisms (Grossman, 2006). The strongest criticisms have come from Marine and Army officers, with retired officers being more candid and dismissive but even serving officers allowing their concerns to come through.

This report is not the place to review the debate, some of which appeared in e-mail exchanges from Lt. Gen. Paul Van Riper (USMC, ret.) to Gen. Pace (USMC), Gen. Hagee (USMC), and GEN Schoomaker (USA), and later between Lt. Gen. Van Riper and Lt. Gen. David Deptula (USAF). Many people in the defense community have read and discussed these exchanges and enjoyed seeing the healthy battle among titans, but if we go beyond the particulars of the exchanges to identify core issues, we see that five are particularly important—and also relevant to the design of high-level military decision-support systems. These are:

1. The clarity and precision with which command principles, decisions, and orders are expressed.
2. Attitudes about uncertainty and the feasibility of achieving control over complex, adaptive, human systems.
3. Optimism about the favorable benefits of technology for the United States in warfare, particularly with regard to information.
4. The levels of command at which EBO-related analysis is most and least appropriate.
5. The balance between centralized control intended to achieve top-to-bottom coherence on extensive delegation and the use of mission orders intended to achieve good adaptation to local events.

Let us discuss the first three. The last two are important and relate to decision-support-system issues such as whether relatively detailed EBO methods are desirable at all echelons of planning (we think not), but they are topics on which we have little to contribute.

Clarity and Precision

After more than a decade of discussion, it is generally agreed now that the "effects" of EBO are at a level between what we might call operational objectives and tasks. The notion is that a commander's operational objectives are often not stated in enough detail to convey a sense of what is needed for good course-of-action development and subsequent operations. EBO, then, has sometimes been described as a new approach that improves upon classic methods of conceiving and expressing decisions and orders. That claim is very much a matter of dispute, with many arguing that in a proper understanding of existing doctrine, the mission, task, and intent are supposed to be conveyed in a single package with readily understood language. In a recent paper, Lieutenant General Paul Van Riper (USMC, ret.) provides a simple example for the case of an operational-level commander developing guidance:

> An analysis by the commander determines that the *center of gravity* for the enemy he faces is a corps size organization. The unit, however, has excellent defenses and the commander decides that a direct attack on it would be very costly. The enemy, though, would be *vulnerable* if attacked while moving which it is likely to do if it sees friendly forces withdrawing. The commander decides to feint a withdrawal. He also decides that the enemy would offer a *critical vulnerability* if attacked as it tried to cross the White River, so he designates the three bridges over that river in his area as *decisive points*. He then makes these bridges objectives and assigns the *mission* of seizing them to one of his own divisions. The unit's missions read, "Seize bridges (*task*) over White River in your zone of action in order to prevent the enemy from continuing to move south (intent)." Finally, he defines the *end-state* he desires: The enemy corps halted north of the White River and damaged to such an extent it will be unable to conduct offensive operations for at least 96 hours, and friendly units in defensible positions south of the river, re-supplied, and prepared to exploit the situation within 6 hours. The *end* is a specified level of damage to the enemy corps. The *means* to accomplish this *end* are the divisions of the friendly corps. The *ways* are the seizure of the three bridges to halt the enemy's movement (Van Riper, 2006 pp. 14–15).

The point of the example is to demonstrate that classic doctrine and language include what many advocates of EBO have claimed were missing—enough description of the commander's intent and rationale so that those being tasked have the knowledge needed for intelligent execution and possible adaptations.[1] Van Riper's example reads clearly and straightforwardly, in terms of both English and logic. In contrast, Figure A.1 shows JFCOM's recent depiction of how objectives, end states, missions, effects, and tasks relate to one another (U.S. Joint Forces Command, 2006a). If one takes the depiction seriously, then confusion and headache are likely. For example, the depiction (1) overrides a useful classic distinction between goals and end state; (2) treats mission as subordinate to objectives at the theater-strategic level, but objectives subordinate to mission at the operational level; (3) intermixes what may broadly be called *goals* with results in the form of *tasks*; and (4) at the tactical level, refers only to mission and tasks, as though the other concepts (e.g., effects and end state) do not apply.

---

[1] See also the Marines' planning manual, which many (including the authors) regard as exceptionally clear on these matters (United States Marine Corps, 2000).

Figure A.1
Proposed Hierarchy of Concepts in a Recent JFCOM Document

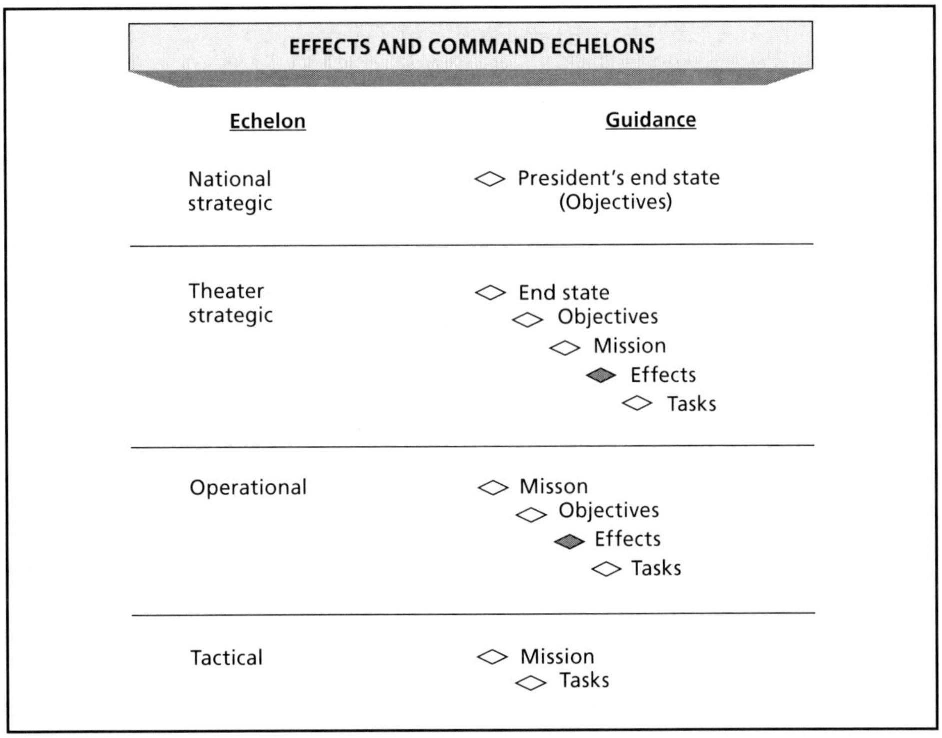

EFFECTS AND COMMAND ECHELONS

Echelon	Guidance
National strategic	◇ President's end state (Objectives)
Theater strategic	◇ End state ◇ Objectives ◇ Mission ◆ Effects ◇ Tasks
Operational	◇ Misson ◇ Objectives ◆ Effects ◇ Tasks
Tactical	◇ Mission ◇ Tasks

RAND *TR422-A.1*

By and large, Van Riper and other critics of EBO are on solid ground when they criticize the convoluted "logic" used in many EBO discussions and briefings.

The relevance of such matters to the current report is that

- A first principle of decision-support systems is that the concepts and structures used must be as clear, logical, and linear as possible.

If this principle is not heeded, it will be substantially more difficult for those a decision-support system is intended to serve to understand issues and make clear, coherent decisions.

Suggesting a final resolution of the semantic difficulties of command and control is beyond the scope of this report. However, we do make the following observation:

- The principle of simplicity and clarity argues for dealing with what are currently being called *effects* in terms of corresponding operational-level *objectives* and *subobjectives*. This would avoid a distinctly unhelpful syntactic nonlinearity.

Other problems exist as well and may be summarized as the *fractal problem*:

- Describing objectives, strategies, tactics, and tasks is a fractal matter—i.e., the concepts apply and are needed at each level, whether that of the president, the theater commander, the Air Force squadron leader, or the Marine platoon leader.
- It follows that efforts to associate these words with one level and not another are doomed, because in the English language, they are used ubiquitously at all levels.
- Efforts over the years to improve the situation by using different words for the different levels (e.g., *mission* versus *objective*) have not worked, in large part because the words are essentially synonyms in the English language.
- The most obvious solution is to accept the fractal nature of the problem and attach level-specifying adjectives or prepositional phrases, as in
  - A Joint Task Force Commander's (JTFC's) mission or a squadron commander's mission.
  - Strategic-level effects or operational-level effects or tactical-level effects.

Attitudes About Uncertainty and Controllability

The second of the five controversial issues relates ultimately to whether uncertainty is seen and treated as a mere annoyance or as something more fundamental. This relates to whether one can reasonably aspire to control developments in some detail or instead aspires to "deal with" the surprises that will assuredly arise.

Two Contrasting Views

It is useful to contrast two views, which can be thought of as the views of air-power enthusiasts and ground-force commanders, respectively, although all contributors to the debate see themselves as contributing to the joint war fight and are by no means as parochial as their critics would have it.

The Air-Power View of Colonel John Warden. The seminal works of Col. John Warden (USAF, ret.) helped shape the Air Force's embrace of effects-based targeting and later EBO (Warden, 1989, 1995). In his widely read 1995 article, he described his system perspective:

> Strategic warfare provides the most positive resolution of conflicts. To execute it well, however, we must reverse our normal method of thinking; we must think from the big to the small, from the top down. We must think in terms of systems; we and our enemies are systems and subsystems with mutual dependencies. Our objective will almost always involve doing something to reduce the effectiveness of the overall system, if you will, to make it more susceptible to the infectious ideas we want to become part of it.
>
> At the same time, we must take necessary action to ensure that the enemy does not do unacceptable damage to our system or any of its subsystems.
>
> We must not start our thinking on war with the tools of war—with the airplanes, tanks, ships, and those who crew them. These tools are important and have their place, but they

cannot be our starting point, nor can we allow ourselves to see them as the essence of war. Fighting is not the essence of war, nor even a desirable part of it. The real essence is doing what is necessary to make the enemy accept our objectives as his objectives.

Warden also proposed a now-famous five-ring model for strategic application of air power (Figure A.2). He proposed not only thinking top-down, as discussed above, but attacking "inside-out," starting with leadership and moving outward, eventually, to fielded forces. His discussion of this concept made it clear that he was thinking predictively about how strategic victory could be achieved with such a strategy, e.g., by decapitating and disconnecting leadership, and so on, to include demoralizing the population. Almost as an afterthought, it seemed, he acknowledged the possible need to attack the adversary's fielded forces, but he treated that possibility as least critical and most likely to be prolonged and bloody. When Warden's approach was taken in Air Force planning for the 1991 war with Iraq (Desert Storm), the original context was indeed an effort to win the war with strategic air power.

Warden's systems perspective was quite apt and survives to this day, but—not surprisingly—many aspects of his original conception have proven to be flawed. In particular, his

Figure A.2
The Warden Five-Ring Model

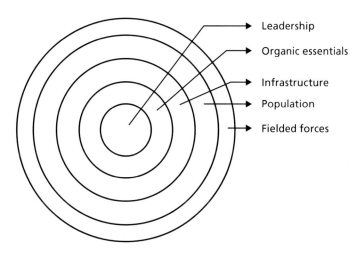

Leadership: Most critical ring comprising decisionmaking entities, command-and-control nodes, etc.

Organic essentials: Second most critical ring comprising those facilities or processes a state requires to survive (power-generation facilities, etc.)

Infrastructure: Third most critical ring comprising the enemy's transportation system (rail lines, bridges, airfields, ports, etc.)

Population: Fourth most critical ring; very difficult to target (morally, internationally) directly; best approached indirectly as North Vietnam did to the United States

Fielded forces: Least critical and most hardened by design; campaigns focusing on this ring tend to be the longest and most bloody

NOTE: Figure adapted from Warden (1989).
RAND *TR422-A.2*

assertion that the key to strategic paralysis consisted of inside-out attack has not been validated and was never a sound basis for a general theory. Instead, it represented a very interesting framework that might sometimes apply but that other times would not. Further, the discussion did not account well for the move toward greater distribution and delegation of authority likely to occur naturally in this networked era, a move that would likely be further accelerated in response to U.S. air power. Such matters were noted by thoughtful inside-the-Air-Force critics (Fadok, 1994), as well as by those who merely responded reflexively against an Air Force theory and the currents of reform and transformation.

A Recently Expressed Ground Commander's View. The frustration of ground-force officers when exposed to EBO planning processes that emphasize efforts to predict developments can be seen in the recent comments of Lt. Gen. James Mattis, head of Marine Corps Combat Development Command and ground commander for the Marines' 2003 offensive in Iraq. As quoted in a news article (Grossman, 2006), Mattis said about EBO that

> You cannot take down a government . . . the same way you can an electrical grid. . . . When you enter into the areas where human beings—with their willpower, their imagination, their courage, their fears, their cultural tendencies—all come to bear, the idea that you can put an algebraic equals sign between something you do and the response that you're going to get is not borne out by the last 5,000 years of human interactions on this planet.

He went on to say, "Watch for these two words: 'predictive analysis' . . . . That'll be the canary in the mineshaft."

Both Sides Are Right, Sometimes

Neither of the contrasting attitudes is absurd in the abstract, even if one accepts that warfare is best studied with the concepts of complex adaptive systems. After all, engineers routinely achieve control over highly complex, dynamic, nonlinear systems, which would show very unstable behaviors but for the measures taken by the engineers to keep the systems away from circumstances of instability (e.g., measures enforcing the envelope of safe performance in an automobile or aircraft). Similarly, throughout history, commanders have systematically prevailed in series of battles despite having only modest intelligence on the details of the enemy's location, movement, and strategy, often doing so by being able to operate within the famous OODA loop of the adversary.[2]

Even though neither side is generally right or wrong, differences can be noted. It seems that Air Force officers are inclined intuitively toward the image of imposing control on a system, whereas ground-force officers are more inclined toward humility and constant reference to the fogs and frictions of war. This accords with their Services' historical experiences and also with the phenomena with which they deal. The sophisticated effects-based targeting used by air forces in recent wars is essentially about physics and relatively static intelligence, whereas even in Operation Iraqi Freedom, U.S. ground forces at battalion levels and below

---

[2] Observe, Orient, Decide, Act. The concept of the OODA loop was introduced by Col. John Boyd (USAF) in the 1970s and is now widely quoted in contexts ranging from air-to-air combat to long-term strategic planning.

were almost as much in the fog of war as their predecessors had been in earlier centuries: They simply did not know what was over the next hill, or whether those they encountered would fight or surrender (Talbot, 2004).

How Do Things Stand on Treatment of Uncertainty?

Against this background, it is reasonable to ask what current doctrinal materials have to say about uncertainty, since those materials form much of the context in which decision-support systems are developed. Our review of documents revealed the following.

The Commander's Handbook on EBO. JFCOM's *Commander's Handbook for an Effects-Based Approach to Joint Operations* says very little about uncertainty, except for the throwaway line, "Thus, JFCs and staff must expect uncertainty and supplement current intelligence with their judgment and intuition" (U.S. Joint Forces Command, 2006a, p. I-2). The handbook discusses risk, but only as something to be mitigated by EBO, i.e., analysis should lead to "a better understanding of risk and the probability of undesired effects" (U.S. Joint Forces Command, 2006a, p. III-14). Although this is a reasonable aspiration, the "spin" reflects a can-do attitude that is inconsistent with that of ground commanders, who are certain only that surprises will occur. The document uses the term *predictive* in a way that many (including us) find worrisome. For example, it states, "JIPB [Joint Intelligence Preparation of the Battlefield], enhanced by an embedded SoSA [Systems of Systems Analysis] approach, produces predictive intelligence with regard to the adversary's probable intent and most likely courses of action" (U.S. Joint Forces Command, 2006a, p. II-I5). Read literally, the statement is true. However, again, the emphasis seems to be on the can-do, rather than the uncertainties of warfare.[3]

The authors of the handbook were not entirely unaware of the issues. For example, they discuss how even best efforts to predict effects will sometimes prove wrong (U.S. Joint Forces Command, 2006a, p. IV-17). Nonetheless, we find the document's discussion to be unnervingly supportive of an approach that takes prediction much too seriously and the near certainty of surprises too lightly. The problem, in our view, is not so much in the literal language, but in the lessons being conveyed.

Evidence of Dissent or a Failure to Coordinate? Curiously, a supplement to JFCOM's handbook (Joint Forces Command, 2006b) does a much better job of discussing uncertainty and unpredictability. It states :

> No wise commander believes that most systems can be understood with anything resembling certainty or that systems can be manipulated with anything approximating deterministic mastery or precision. . . . Most . . . will confound detailed understanding: their nodes and links often cannot be accurately mapped. . . . Systems will often exhibit unpredictable, surprising, and uncontrollable behaviors. . . . Effects-based approach calls for a

---

[3] Other sources place emphasis similarly. For example, the JFACC handbook (Air Combat Command (ACC/A3C), 2005) exhorts building flexibility into plans, e.g., scheduling on-call close air support and air interdiction packages (which only a decade ago was controversial) (p. 21); overall, however, it also is short on discussion of uncertainty. Air Force basic doctrine acknowledges (p. 15) that war is a complex and chaotic human endeavor with fog and friction. It mentions that an enemy can be highly unpredictable and even irrational.

significant level of humility. . . . *This concept argues rather for a framework that sees operations as learning—that is, military actions themselves become an experiential means of learning about the target system* [emphasis added] (p. Supplement-n5).

This is the type of language that we believe should be front and center, not relegated to footnotes or a supplement subtitled "Theory."

More Caution by the Top Leadership. Interestingly, some of the most senior proponents of EBO also have strong cautionary language. For example, in 2001, then-Maj. Gen. Deptula wrote (Deptula, 2001, p. 15):

> Intelligence is never complete about an enemy. No intelligence system will ever fully comprehend adversary strategic centers of gravity, the constituent operational systems, and the set of individual targets making up each system. Moreover, an enemy will attempt to negate the effects of actions taken against them while trying to respond effectively. As a consequence, parallel war conducted to achieve rapid decisive operations may involve more than one set of force application, even if the resources are available to attack all the known elements of the identified systems that might affect the enemy. Any enemy may react to attack in ways not anticipated, may have elements unknown to the friendlies, or the friendlies may not possess the capacity to quickly and effectively counter an enemy move. Any or all of these contingencies may change the calculus of the original parallel attack formula requiring additional application of force and lengthening the time to achieve desired effects.

Good Language in Conceiving the Commander's Predictive Environment. Another document we looked at was a recent Broad Area Announcement related to the Commander's Predictive Environment. Despite the name, the thrust seems to us unexceptionable (Broad Area Announcement (BAA) #06-07-IKFA, November 22, 2005):

> The objective of the Commander's Predictive Environment (CPE) program is to provide a decision support environment that enables the Joint Force Commander/Joint Force Air Component Commander (JFC/JFACC) to anticipate and shape the future battlespace. As envisioned, CPE will have four key capabilities: 1) Understand the Battlespace, 2) Anticipate Plausible Futures, 3) Evaluate Courses of Action, and 4) Access and Share Battlespace Information. This BAA specifically addresses the first key capability, "Understand the Battlespace."

It may be that the authors had in mind a higher degree of confidence in being able to understand and anticipate than we have, but at least the language of the Broad Area Announcement does no harm.

Recommendations

The implications of this controversy about uncertainty, for this report, seem to us as follows:

- High-level decision-support systems for military commanders such as JFACCs or JTFCs should not be tilted toward one or the other perspective, since context drives everything and neither perspective represents a sound general theory.
- Although many effects can plausibly be predicted with a considerable degree of confidence (e.g., those associated with disruption of electrical power or the ability of ground armies to maneuver in the open), others are inherently much more uncertain—even as to whether particular actions will prove productive or counterproductive. As a result, different attitudes about uncertainty are needed for assessing the issues, and different tools may well be necessary as well.
- By and large, we see greater value for computer models and related "analytical" decision-support systems when dealing with more nearly predictable phenomena and greater value for human-intensive methods, such as gaming (including the foresight methods of Chapter Four), to plan for dealing with and adapting to more-uncertain phenomena.

Optimism About Technology and Information Systems

Few issues separate senior military officers and theorists more than that of information dominance. In the history of warfare, Joint Vision 2010 and its successor Joint Vision 2020[4] will probably be seen as having been in many ways prescient and appropriately ambitious. This work laid the basis for much of the transformation that has been occurring in U.S. military forces, at least with regard to high-end warfare.

One aspect of the vision has been information dominance, and indeed the United States has made profound leaps in being able to achieve a shared situational understanding among commanders and a quite remarkable high-level understanding of operational-level forces and activities in large-scale combat.

Unfortunately, but precisely as was predictable, adversaries of the United States have turned to forms of warfare that undercut our technological advantages. This is now a commonplace, but it is worth reviewing because of its relevance to decision support. Consider in particular the enormous emphasis U.S. doctrinal materials and professional writing place on achieving information dominance. The aspiration and emphasis is laudable, but the assertiveness of the language and lack of critical thinking are very troubling. Large exercises have been held in which Blue is *assumed* to have essentially perfect information and in which the adversary fights to Blue's strengths. There has been consistent failure to discuss prominently and seriously the need to prepare for circumstances in which the United States does *not* have information dominance.

---

[4] Joint Vision 2010, published in the mid-1990s, was the initial conceptual template for how America's armed forces would channel the vitality of their people and leverage technological opportunities to achieve new levels of effectiveness in joint war-fighting. Joint Vision 2020 builds upon and extends Joint Vision 2010 to guide the continuing transformation of the armed forces (Director for Strategic Plans and Policy, 2000).

Figure A.3, adapted from a contribution of the senior author to an earlier study on network-centric operations (National Research Council, 2000), illustrates the problem. If we try to sharpen our definition of what information dominance *means*, we soon recognize that what matters is not the inputs (e.g., the number of surveillance systems) or even the intermediate outputs (e.g., the area searched per day), but rather the ultimate outputs, such as whether Red and Blue have the information they need to pursue effectively their chosen strategies. The strength of Blue surveillance may limit what Red strategies can be, which is good, but given that Red adopts appropriate strategies, the United States can readily find itself in the wrong quadrant of the chart. Crudely speaking, Iraq in 2006 is analogous to the Vietnam war in that the United States has superiority in technical information collection but lacks the information it needs. In contrast, the adversary is able to pursue his strategy with the on-the-ground information available to him.

The implication for high-level decision support, such as that needed by a JFACC, is that

- Decision-support systems should not be infected by the widespread tendency to express visions and idealizations as though they were achievable realities, or even current realities.

Figure A.3
Information Balance in Conflict

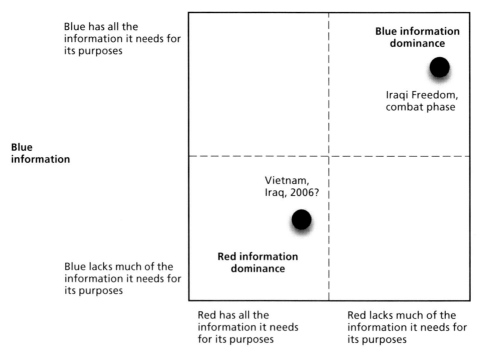

- The language of decision-support systems (including the metrics used for modeling and simulation outputs) should be scrupulously reviewed to focus on functionally significant metrics, such as those suggested in Figure A.2.

We could not pursue these issues in any detail in the current study, but two examples based on real-world problems may make the larger point. Given a desire by policymakers to have persistent surveillance, will analysis and decision-support systems generate data that conveniently define "persistent" so that the results are favorable, or will they generate data that relate more to the ultimate mission, such as the probability of detecting, tracking, and killing critical mobile targets? Similarly, given the desire for global-strike capability, will analysis and decision-support systems implicitly assume that the intelligence of the time and place to be struck are correct, or will they highlight uncertainties? During Operation Iraqi Freedom, the physical aspects of strike operations were accomplished superbly, but the percentage of success in leadership attacks was almost vanishingly small—*even though those approving, directing, and executing the strikes were relatively confident!*

Bibliography

Air Combat Command (ACC/A3C), *Air Force Air and Space Operations Center (AOC) Operating Concept*, Washington, D.C., 2005.

Allen, Patrick, and Barry Wilson, *The Secondary Land Theater Model*, Santa Monica, Calif.: RAND Corporation, N-2625-NA, 1987. Online at:
http://www.rand.org/pubs/notes/N2625

The American Heritage® Dictionary of the English Language, 4th ed., Boston, Mass.: Houghton Mifflin, 2003.

Bankston, LTC Boyd (USA), and LTC Todd Key (USA), "White Paper on Capabilities-Based Planning (Draft)," in *Proceedings of MORS: Capabilities Based Planning II*, McLean, Va.: Military Operations Research Society, April 4–6, 2006. As of April 11, 2007:
http://www.mors.org/meetings/cbp_II/briefs/bankston_key.pdf

Bazerman, Max H., *Managerial Decision Making*, 2nd ed., New York: John Wiley, 1986.

Beagle, T. W., *Effects-Based Targeting: Another Empty Promise?* Maxwell AFB, Ala.: Air University Press, 2000.

Botterman, Maarten, Jonathan Cave, James P. Kahan, Neil Robinson, Rebecca Shoob, Robert Thomson, and Lorenzo Valeri, *Foresight Cyber Trust and Crime Prevention Project: Gaining Insight from Three Different Futures*, London: Office of Science and Technology, 2004.

Burdick, Chuck, et al., "Plan Building in JWARS," in *Proceedings of 11th Conference on Computer Generated Forces and Behavioral Representation*, 2002, pp. 421–430.

Clark, General Wesley K. (USA, ret.), *Waging Modern War,* Cambridge, Mass.: Public Affairs, Perseus Books, 2001.

Cohen, William, *Report of the Quadrennial Defense Review*, Washington, D.C.: Department of Defense, 1997.

Crane, Conrad C., "Beware of Boldness," *Parameters*, Vol. XXXVI, No. 2, Summer 2006, pp. 88–97.

Davis, Paul K., "Planning for Adaptiveness," in *New Challenges for Defense Planning: Rethinking How Much Is Enough*, Paul K. Davis (ed.), Santa Monica, Calif.: RAND Corporation, MR-400-RC, 1994, pp. 51–73. Online at:
http://www.rand.org/pubs/monograph_reports/MR400/index.html

———, *Effects-Based Operations (EBO): A Grand Challenge for the Analytical Community*, Santa Monica, Calif.: RAND Corporation, MR-1477-USJFCOM/AF, 2001. Online at:
http://www.rand.org/pubs/monograph_reports/MR1477

———, *Analytic Architecture for Capabilities-Based Planning, Mission-System Analysis, and Transformation*, Santa Monica, Calif.: RAND Corporation, MR-1513-OSD, 2002a. Online at:
http://www.rand.org/pubs/monograph_reports/MR1513

———, "Synthetic Cognitive Modeling of Adversaries for Effects-Based Planning," *Proceedings of the SPIE*, Vol. 4716, No. 27, 2002b, pp. 236–250.

———, "Uncertainty-Sensitive Planning," in Stuart Johnson, Martin Libicki, and Gregory Treverton (eds.), *New Challenges, New Tools for Defense Decisionmaking*, Santa Monica, Calif.: RAND Corporation, MR-1576-RC, 2003. Online at:
http://www.rand.org/pubs/monograph_reports/MR1576

———, "Rethinking Families of Models," *Proceedings of SPIE, Enabling Technologies for Simulation Science VIII*, Dawn A. Trevisani and Alex F. Sisti (eds.), Vol. 5423, 2004, pp. 17–31.

———, "New Paradigms and New Challenges," keynote presentation, Military Track, *Proceedings of the Winter Simulation Conference*, 2005a.

———, "Introduction to Multiresolution, Multiperspective Modeling (MRMPM) and Exploratory Analysis," in *A Tutorial on "New" Analysis Techniques: Understanding & Applications: Presentations*, McLean, Va.: Military Operations Research Society, February 18, 2005b. As of April 11, 2007:
http://www.mors.org/meetings/tutorial/tutorial_presentations.htm

Davis, Paul K., and John Arquilla, *Thinking About Opponent Behavior in Crisis and Conflict: A Generic Model for Analysis and Group Discussion*, Santa Monica, Calif.: RAND Corporation, N-3322-JS, 1992. Online at:
http://www.rand.org/pubs/notes/N3322.

Davis, Paul K., Steven C. Bankes, and Michael Egner, *Enhancing Strategic Planning with Massive Scenario Generation: Theory and Experiments*, Santa Monica, Calif.: RAND Corporation, TR-392-OSD, 2006. Online at:
http://www.rand.org/pubs/technical_reports/TR392

Davis, Paul K., and James H. Bigelow, *Motivated Metamodels: Synthesis of Cause-Effect Reasoning and Statistical Metamodeling*, Santa Monica, Calif.: RAND Corporation, MR-1570-AF, 2003. Online at:
http://www.rand.org/pubs/monograph_reports/MR1570

Davis, Paul K., James H. Bigelow, and Jimmie McEver, *Exploratory Analysis and a Case History of Multiresolution, Multiperspective Modeling*, Santa Monica, Calif.: RAND Corporation, RP-925, 2001. Online at:
http://www.rand.org/pubs/reprints/RP925

———, *Effects of Terrain, Maneuver Tactics, and C4ISR on the Effectiveness of Long-Range Precision Fires: A Stochastic Multiresolution Model (PEM) Calibrated to High-Resolution Simulation*, Santa Monica, Calif.: RAND Corporation, MR-1138-OSD, 2000. Online at:
http://www.rand.org/pubs/monograph_reports/MR1138

Davis, Paul K., James Bonomo, Henry Willis, and Paul Dreyer, "Analytic Tools for Strategic Planning and Investment in the Missile Defense Agency," in *Proceedings of 3d Annual U.S. Missile Defense Conference*, 2005.

Davis, Paul K., Jonathan Kulick, and Michael Egner, *Implications of Modern Decision Science for Military Decision-Support Systems*, Santa Monica, Calif.: RAND Corporation, MG-360-AF, 2005. Online at:
http://www.rand.org/pubs/monographs/MG360

Davis, Paul K., and Lou Finch, *Defense Planning in the Post Cold-War Era: Giving Meaning to Flexibility, Adaptiveness, and Robustness of Capability*, Santa Monica, Calif.: RAND Corporation, MR-322-JS, 1993. Online at:
http://www.rand.org/pubs/monograph_reports/MR322

Davis, Paul K., David Gompert, and Richard Kugler, *Adaptiveness in National Defense: The Basis of a New Framework*, Santa Monica, Calif.: RAND Corporation, IP-155, 1996. Online at:
http://www.rand.org/pubs/issue_papers/IP155

Davis, Paul K., and Amy Henninger, *Analysis, Analysis Practices, and Implications for Modeling and Simulation*, Santa Monica, Calif.: RAND Corporation, OP-176-OSD, 2007. Online at:
http://www.rand.org/pubs/occasional_papers/OP176

Davis, Paul K., and James P. Kahan, "Theory and Methods for Supporting High-Level Decision Making," in *Proceedings of SPIE: Enabling Technologies for Simulation Science X*, Vol. 6227, May 2006.

Davis, Paul K., Jimmie McEver, and Barry Wilson, *Measuring Interdiction Capabilities in the Presence of Anti-Access Strategies: Exploratory Analysis to Inform Adaptive Strategies for the Persian Gulf*, Santa Monica, Calif.: RAND Corporation, MR-1471-AF, 2002. Online at: http://www.rand.org/pubs/monograph_reports/MR1471

Davis, Paul K., and Russell D. Shaver, "An Analytic Framework for Capability Reviews and Tradeoffs Among Options," Discussion, unpublished RAND research, 2006.

Dawes, Robyn M., "The Robust Beauty of Improper Linear Models," *American Psychologist*, Vol. 34, 1979, pp. 571–582.

Defense Science Board, *Phase I Report of the Defense Science Task Force on Joint Experimentation*, Washington, D.C.: Office of the Under Secretary of Defense for Acquisition, Technology, and Logistics, 2003a.

———, *The Role and Status of Red Teaming Activities*, Washington, D.C.: Office of the Under Secretary of Defense for Acquisition, Technology, and Logistics, 2003b.

Department of Defense, "DoD News Briefing—Secretary Rumsfeld and Gen. Pace (August 20, 2002)." As of April 11, 2007: http://www.defenselink.mil/transcripts/2002/t08202002_t0820sd.html

Deptula, Brigadier General David, *Effects-Based Operations: Change in the Nature of Warfare,* Arlington, Va.: Aerospace Education Foundation, 2001. As of April 11, 2007: http://www.afa.org/media/reports/

Dewar, James, *Assumption Based Planning*, Cambridge, UK: Cambridge University Press, 2003.

Director for Strategic Plans and Policy, Strategy Division, *Joint Vision 2020*, Washington, D.C.: U.S. Government Printing Office, June 2000.

Dobbins, James, John G. McGinn, Keith Crane, Seth G. Jones, Rollie Lal, Andrew Rathmell, Rachel M. Swanger, and Anga Timilsina, *America's Role in Nation-Building: From Germany to Iraq,* Santa Monica, Calif.: RAND Corporation, MR-1753-RC, 2003. Online at: http://www.rand.org/pubs/monograph_reports/MR1753

Dreyer, Paul, and Paul K. Davis, *A Portfolio-Analysis Tool for Missile Defense (PAT-MD): Methodology and User's Manual,* Santa Monica, Calif.: RAND Corporation, TR-262-MDA, 2005. Online at: http://www.rand.org/pubs/technical_reports/TR262

Dunnigan, James F., *How to Make War: A Comprehensive Guide to Modern Warfare in the Twenty-First Century*, 4th ed., New York: Harper Paperbacks, 2003.

Dupuy, Trevor Nevitt, *Understanding War: History and Theory of Combat*, New York: Paragon House, 1987.

Fabozzi, Frank J., and Harry M. Markowitz, *The Theory and Practice of Investment Management*, New York: Wiley, 2002.

Fadok, David S., *John Boyd and John Warden: Air Power's Quest for Strategic Paralysis*, Maxwell AFB, Ala.: Air University, 1994.

Fox, Daniel, "Using Exploratory Modeling," in Stuart Johnson, Martin Libicki, and Gregory F. Treverton (eds.), *New Challenges, New Tools for Defense Decisionmaking*, Santa Monica, Calif.: RAND Corporation, MR-1576-RC, 2003, pp. 258–298. Online at: http://www.rand.org/pubs/monograph_reports/MR1576

Franks, General Tommy (USA), with Malcolm McConnell, *American Soldier*, New York: Regan Books, 2004.

Gigerenzer, Gerd and Reinhard Selten, *Bounded Rationality: The Adaptive Toolbox*, Cambridge, Mass.: MIT Press, 2002.

Gilovich, Thomas, Dale Griffin, and Daniel Kahneman (eds.), *Heuristics and Biases: The Psychology of Intuitive Judgment*, Cambridge, UK: Cambridge University Press, 2002.

Gladwell, Malcolm, *Blink: The Power of Thinking Without Thinking*, London, UK: Little Brown, 2005.

Gompert, David, Irving Lachow, and Justin Perkins, *Battle-Wise: Gaining Advantage in Networked Warfare*, Washington, D.C.: National Defense University, 2005.

Gordon, Michael R., and Bernard Trainor, *Cobra II: The Inside Story of the Invasion and Occupation of Iraq*, New York: Pantheon, 2006.

Grossman, Elaine M., "Effects-Based Operations Under Fire: A Top Commander Acts to Defuse Military Angst on Combat Approach," *Inside the Pentagon*, April 20, 2006.

Haimes, Yacov, *Risk Modeling, Assessment, and Management*, 2nd ed., New York: John Wiley & Sons, 2004.

Helmer-Hirschberg, Olaf, *Analysis of the Future: The Delphi Method*, Santa Monica, Calif.: RAND Corporation, P-3558, 1967. Online at: http://www.rand.org/pubs/papers/P3558

Henry, Ryan, "Capabilities Based Planning II Workshop (Keynote): Results of the 2006 WDFR," in *Proceedings of MORS: Capabilities Based Planning II*, Alexandria, Va.: Military Operations Research Society, April 4–6, 2006. Online at: http://www.mors.org/meetings/cbp_II/briefs/henry.pdf

Hillestad, Richard, and Paul K. Davis, *Resource Allocation for the New Defense Strategy: The DynaRank Decision-Support System*, Santa Monica, Calif.: RAND Corporation, MR-996-OSD, 1998. Online at: http://www.rand.org/pubs/monograph_reports/MR996

Hitch, Charles J., and Roland N. McKean, *Economics of Defense in the Nuclear Age*, New York: Scribner, 1965 (published originally by the RAND Corporation and Harvard University Press, 1960).

Hoffman, Hugh F.T., "Capabilities Packaging in Adaptive Planning Overview," in *Proceedings of MORS: Capabilities Based Planning II*, Alexandria, Va.: Operations Research Society, April 4–6, 2006. Online at: http://www.mors.org/meetings/cbp_II/briefs/hoffman.pdf

Horne, Gary E., and Mary Leonardi (eds.), *Maneuver Warfare Science*, Quantico, Va.: U.S. Marine Corps Combat Development Command, 2001.

Ilachinski, Andrew, *Artificial War: Multiagent-Based Simulation of Combat*, Hackensack, N.J.: World Scientific Publishing Company, 2004.

Jobbagy, Zoltan, *Theory, Reality and the Nature of War: Effects-Based Operations Meet Post-Conflict Operations*, Clingendael, The Netherlands: Centre for Strategic Studies, 2006.

Jones, Carl, *JICM 3.5 "J" Language*, Santa Monica, Calif.: RAND Corporation, Discussion, unpublished RAND research, 1999.

Kahan, James P., and Paul K. Davis, "Foresight for Commanders: A Methodology to Assist Planning for Effects-Based Operations," in *Proceedings of SPIE*, 2006.

Kahan, James P., D. Robert Worley, and Cathleen Stasz, *Understanding Commander's Information Needs*, Santa Monica, Calif.: RAND Corporation, R-3761-1-A, 2000. Online at: http://www.rand.org/pubs/reports/R3761-1

Kahneman, Daniel, "Maps of Bounded Rationality: A Perspective on Intuitive Judgment and Choice (Nobel Prize Lecture)," Stockholm University, Stockholm, Sweden, December 8, 2002. As of April 11, 2007: http://www.nobel.se/economics/laureates/2002/kahneman-lecture.html

Kahneman, Daniel, and Amos Tversky, "Prospect Theory: An Analysis of Decision Under Risk," *Econometrica*, Vol. 47, 1979, pp. 263–291.

Keaney, Thomas A., and Eliot A. Cohen, *Gulf War Air Power Survey Summary Report*, Washington, D.C.: Department of the Air Force, 1993.

Keegan, John, *The Face of Battle: A Study of Agincourt, Waterloo, and the Somme*, New York: Penguin (Non-Classics), 1983.

Keeney, R., and Howard Raiffa, *Decisions with Multiple Objectives: Preferences and Value Tradeoffs*: New York, Cambridge University Press, 1993 (first published in 1976 by John Wiley & Sons).

Kent, Glenn A., and David A. Ochmanek, *A Framework for Modernization Within the United States Air Force*, Santa Monica, Calif.: RAND Corporation, MR-1706-AF, 2003. Online at: http://www.rand.org/pubs/monograph_reports/MR1706/index.html

Klein, Gary, *Sources of Power: How People Make Decisions*, Cambridge, Mass.: MIT Press, 1998.

Knight, Frank Hyneman, *Risk, Uncertainty, and Profit*, Boston, Mass.: Houghton-Mifflin, 1921.

Lambeth, Benjamin, *Desert Storm and Its Meaning: The View from Moscow*, Santa Monica, Calif.: RAND Corporation, R-4164-AF, 1992. Online at: http://www.rand.org/pubs/reports/R4164

———, *NATO's Air War for Kosovo: A Strategic and Operational Assessment,* Santa Monica, Calif.: RAND Corporation, MR-1365-AF, 2001. Online at: http://www.rand.org/pubs/monograph_reports/MR1365/index.html

Lempert, Robert J., Steven W. Popper, and Steven C. Bankes, *Shaping the Next One Hundred Years: New Methods for Quantitative, Long-Term Policy Analysis,* Santa Monica, Calif.: RAND Corporation, MR-1626-RPC, 2003. Online at: http://www.rand.org/pubs/monograph_reports/MR1626

Light, Paul C., *The Four Pillars of High Performance*, New York: McGraw-Hill, 2004.

Linstone, Harold A., and Murray Turoff (eds.), *The Delphi Method: Techniques and Applications,* Newark, N.J.: New Jersey Institute of Technology, 2002. Online at: http://www.is.njit.edu/pubs/delphibook/delphibook.pdf

March, James G., *A Primer on Decision Making: How Decisions Happen*, New York: The Free Press, 1994.

McCrabb, Maris, *Explaining Effects*, Rome, NY: Air Force Research Laboratory Information Directorate, 2001.

Molander, Roger, Peter Wilson, David Mussington, and Richard Mesic, *Strategic Information Warfare Rising*, Santa Monica, Calif.: RAND Corporation, MR-964-OSD, 1998. Online at: http://www.rand.org/pubs/monograph_reports/MR964

Morgan, M. Granger, and Max Henrion, *Uncertainty: A Guide to Dealing with Uncertainty in Quantitative Risk and Policy Analysis*, New York: Cambridge University Press, 1990.

Murdock, Clark A., *The Role of Red Teaming in Defense Planning*, Arlington, Va.: Hicks and Associates, Inc., 2003.

Murray, Williamson, "Red Teaming: Its Contribution to Past Military Effectiveness," *Defense Adaptive Red Team (DART) Report*, Washington, D.C.: Office of the Under Secretary of Defense for Acquisition, Technology, and Logistics, 2002.

———, *Thoughts on Red Teaming*, Arlington, Va.: Hicks and Associates, Inc, 2003.

Mussington, David, "The 'Day After': Methodology and National Security Analysis," in Stuart Johnson, Martin Libicki, and Gregory Treverton (eds.), *New Challenges, New Tools for Defense Decisionmaking*, Santa Monica, Calif.: RAND Corporation, MR-1576-RC, 2003. Online at: http://www.rand.org/pubs/monograph_reports/MR1576

National Research Council, *Network Centric Naval Forces*, Washington, D.C.: Naval Studies Board, National Academies Press, 2000.

———, *The Role of Experimentation in Building Future Naval Forces*, Washington, D.C.: National Academies Press, 2004.

———, *Naval Analytical Capabilities: Improving Capabilities-Based Planning*, Washington, D.C.: National Academies Press, 2005.

————, *Defense Modeling, Simulation, and Analysis: Meeting the Challenge,* Washington, D.C.: National Academies Press, 2006 .

Neustadt, Richard E., and Ernest R. May, *Thinking in Time: The Uses of History for Decision Makers*, New York: The Free Press, 1986.

Ochmanek, David A., Edward Harshberger, David E. Thaler, and Glenn A. Kent, *To Find, and Not to Yield: How Advances in Information and Firepower Can Transform Theater Warfare*, Santa Monica, Calif.: RAND Corporation, MR-958-AF, 1998. Online at: http://www.rand.org/pubs/monograph_reports/MR958

Pape, Robert A., *Bombing to Win: Air Power and Coercion in War*, Ithaca, N.Y.: Cornell University Press, 1996.

Parnell, Greg, "Value-Focused Thinking Using Multiple Objective Decision Analysis," in *Methods for Conducting Military Operational Analysis: Best Practices in Use Throughout the Department of Defense*, McLean, Va.: Military Operations Research Society, June 2005.

Pirnie, Bruce, and Sam B. Gardiner, *An Objectives-Based Approach to Military Campaign Analysis*, Santa Monica, Calif.: RAND Corporation, MR-656-JS, 1996. Online at: http://www.rand.org/pubs/monograph_reports/MR656

Prieutula, Michael J., Kathleen M. Carley, and Les Gasser (eds.), *Simulating Organizations: Computational Models of Institutions and Groups*, Menlo Park, Calif.: AAAI Press/MIT Press, 1998.

Raiffa, Howard, *Decision Analysis: Introductory Lectures on Choices Under Uncertainty,* Reading, Mass.: Addison-Wesley, 1968.

Rayburn, Major General Bentley (USAF), "Effects-Based Assessment," briefing, 2006.

Rosen, Julie A., and Wayne L. Smith, "Influence Net Modeling with Causal Strengths: An Evolutionary Approach," in *Proceedings of the Command and Control Research Symposium*, Monterey, Calif., 1996.

Rumsfeld, Donald, *Quadrennial Defense Review Report*, Washington, D.C.: Department of Defense, 2001.

————, *Quadrennial Defense Review Report*, Washington, D.C.: Department of Defense, 2006.

Saaty, Thomas L., *Decision Making for Leaders: The Analytic Hierarchy Process for Decisions in a Complex World*, New Edition 2001, Analytic Hierarchy Process Series, Vol. 2, Pittsburgh, Penn.: RWS Publications, 1999.

Sanchez, Susan M., and Thomas W. Lucas, "Exploring the World of Agent-Based Simulations: Simple Models, Complex Analysis," *Proceedings of the Winter Simulation Conference*, 2000.

Santos, Eugene, and Qunhua Zhao, "Adversarial Models for Opponent Intent Inferencing," in A. Kott and W. McEneaney (eds.), *Adversarial Reasoning: Computational Approaches to Reading the Opponent's Mind*, Boca Raton, Fla.: CRC Press, 2006, pp. 1–22.

Schwabe, William, and Barry Wilson, *Analytic War Plans: Adaptive Force-Employment Logic in the RAND Strategy Assessment System (RSAS)*, Santa Monica, Calif.: RAND Corporation, N-3051-NA, 1990. Online at: http://www.rand.org/pubs/notes/N3051

Schwartz, Peter, *The Art of the Long View: Planning for the Future in an Uncertain World*, New York: Currency, 1995.

Simulex, "Synthetic Environments for Analysis and Simulation (SEAS)," West Lafayette, Ind., 2006. As of April 11, 2007: http://www.simulexinc.com/products/technology/

Sinnreich, Richard Hart, *Red Team Insights from Army Wargaming*, Arlington, Va.: Hicks and Associates, Inc., 2002.

Smith, Edward A., *Effects Based Operations: Applying Network Centric Warfare in Peace, Crisis, and War*, Washington, D.C.: Command and Control Research Program (CCRP), Office of the Assistant Secretary of Defense, 2003.

———, *Complexity, Networking, and Effects-Based Operations*, Washington, D.C.: Command and Control Research Program (CCRP), Office of the Assistant Secretary of Defense, 2006.

Surowiecki, James, *The Wisdom of Crowds*, New York: Anchor Books, 2005.

Talbot, David, "How Technology Failed in Iraq," *Technology Review*, November 2004.

Tilson, John, MG Waldo D. Freeman (USA, ret.), CAPT William R. Burns (USN, ret.), Lt.Col John E. Michel (USAF), COL Jack A. LeCuyer (USA, ret.), MG Robert H. Scales (USA, ret.), and D. Robert Worley, *Learning to Adapt to Asymmetric Threats*, Alexandria, Va.: Institute for Defense Analyses, D-3114, 2005.

Uhrmacher, Adelinde, and William Swartout, "Agent-Oriented Simulation," in Mohammad S. Obaidet and Georgios Papadimitrious (eds.), *Applied System Simulation*, Dordrecht, Netherlands: Kluwer Academic, 2003, pp. 239–259.

United States Air Force, *Air Force Basic Doctrine*, 2003.

United States Marine Corps, *Marine Corps Planning Process*, MCWP 5-1, Marine Corps Combat Development Command, 2000.

U.S. Air Force Air Materiel Command, "Commander's Predictive Environment: Understand the Battlespace," Rome, N.Y., updated August 18, 2006. As of April 11, 2007:
http://www1.fbo.gov/spg/USAF/AFMC/AFRLRRS/Reference%2DNumber%2DBAA%2D06%2D07%2DIFKA/SynopsisP.html

U.S. Joint Forces Command, Joint Warfighting Center, *Commander's Handbook for an Effects-Based Approach to Joint Operations*, Suffolk, Va.: Joint Forces Command, 2006a.

———, *Supplement to Commander's Handbook for an Effects-Based Approach to Joint Operations (Theory)*, Suffolk, Va.: Joint Forces Command, 2006b.

van de Riet, O.A.W.T., M. van het Loo, and J.P. Kahan, "Developing Scenarios to Support Interactive Policy Analysis: Experiences from Two Policy Analysis Studies," *International Journal of Technology, Policy and Management*, Vol. 5, No. 2, 2005, pp. 132–145.

Van Riper, Lieutenant General Paul K. (USMC, ret.), *Planning for and Applying Military Force: An Examination of Terms*, Carlisle, Penn.: Strategic Studies Institute of the U.S. Army War College, 2006.

Veit, Clairice T., Monti Callero, and Barbara Rose, *Introduction to the Subjective Transfer Approach to Analyzing Systems*, Santa Monica, Calif.: RAND Corporation, R-3021-AF, 1984. Online at: http://www.rand.org/pubs/reports/R3021

Wagenhals, Lee, Insub Shin, and Alexander E. Levis, "Executable Models of Influence Nets Using Design/CPN," briefing, 2001.

Warden, Col. John A., III (USAF), "The Enemy as a System," *Airpower Journal*, Spring 1995.

———, *The Air Campaign: Planning for Combat*, Washington, D.C.: Pergamon Brassey's, 1989.

Warner, Edward L., and Glenn A. Kent, *A Framework for Planning the Employment of Air Power in Theater War*, Santa Monica, Calif.: RAND Corporation, N-2038-AF, 1984. Online at: http://www.rand.org/pubs/notes/N2038

Woodward, Bob, *Plan of Attack*, New York: Simon & Schuster, 2004.